Workflows for Climate Change Adaptation Analytics

Fernando DePaolis

Workflows for Climate Change Adaptation Analytics

A Practitioner's Handbook with Applications in R

Springer

Fernando DePaolis
Environmental Policy and Management
Program
Middlebury Institute of International
Studies at Monterey
Monterey, CA, USA

ISBN 978-3-032-14003-6 ISBN 978-3-032-14004-3 (eBook)
https://doi.org/10.1007/978-3-032-14004-3

This Springer imprint is published by the registered company Springer Nature Switzerland AG
The registered company address is: Gewerbestrasse 11, 6330 Cham, Switzerland

To Maruja, Veronica, Clara, Umberto, and Julia

Preface

Over the course of more than 25 years teaching in graduate programs focused on quantitative policy analysis, regional economics, and—more recently—environmental planning, and sustainability, I have had the privilege of working with hundreds of bright, motivated students—many of whom are now practicing professionals engaged in critical efforts to address climate change and build resilience in their communities. From municipal planners and environmental consultants to regional transportation analysts and nonprofit advocates, these individuals bring deep domain knowledge, strong commitment to the public good, and an ability to think systemically about complex, interconnected problems.

What many of them share, however, is a persistent unease when it comes to applying quantitative analysis in practice. Despite having taken courses in statistics, research methods, or perhaps even GIS and introductory data science, they often find themselves challenged when faced with real-world datasets, ill-defined problems, and the need to deliver actionable insights to decision-makers. This is not a shortcoming on their part—it is a reflection of a broader gap between the way quantitative methods are traditionally taught and the messy, uncertain, fast-moving environments in which climate adaptation decisions are actually made.

This book was written to bridge that gap.

Workflows for Climate Change Adaptation Analytics: A Practitioner's Handbook with Applications in R is intended for professionals who need to use data in the service of climate adaptation but who are not trained statisticians or full-time data scientists. These are the people designing transportation plans that must account for future flood risks, managing conservation programs that respond to shifting ecological baselines, or developing housing and land use policies in regions facing heatwaves, droughts, or sea level rise. They understand the importance of evidence-based decisions—but they need guidance on how to turn raw data into meaningful, credible, and interpretable analysis.

What sets this handbook apart is its focus on workflows—a term I use intentionally. In professional practice, it is not enough to know a method in isolation. Practitioners need a repeatable, flexible process that connects the dots: defining the question, preparing the data, selecting the right methods, executing the analysis, interpreting the results, and communicating them effectively. Each workflow presented in this book is designed to do just that, using R as the computational platform, not because it is the only choice, but because it is powerful, open-source, and widely adopted in the scientific and policy communities.

The book is organized around five key areas of analytical practice that are particularly relevant to climate adaptation work:

1. **Descriptive statistics and visualization**—The foundation for any analysis, helping practitioners understand what their data can tell them through summaries, distributions, and visual representation.
2. **Regression analysis**—A versatile tool for examining relationships between variables, testing hypotheses, and generating predictive insights.
3. **Time series analysis**—Essential for understanding how environmental and social indicators change over time, and for detecting trends and seasonality in climate-sensitive variables.
4. **Spatial analysis**—Critical for evaluating the geographic dimensions of climate impacts and for integrating locational context into planning decisions.
5. **Non-parametric techniques**—Useful when data do not meet the assumptions of classical statistical models, offering robust alternatives that often align better with the nature of real-world datasets, especially those generated by surveys and opinion polls.

Each chapter includes an applied example, based on publicly available or realistically simulated data, that walks through a complete analysis from start to finish. Along the way, I include code snippets written in R, annotations explaining what the code is doing and why, and a discussion of how to interpret the outputs in a planning or policy context. The goal is not to be exhaustive, but to be useful—to give readers enough knowledge and confidence to adapt the workflows to their own needs and situations.

This is not a textbook in the traditional sense. You won't find long proofs, heavy mathematical derivations, or extended discussions of statistical theory. There are many excellent resources for that, and I encourage readers to consult them as their needs grow. What you will find here is something more practical: a field guide for working professionals who want to put data to work in the service of climate resilience and adaptation.

Throughout the book, I have tried to maintain a conversational tone, much like the one I use in the classroom. My intention is to be a guide, not a gatekeeper. The audience I have in mind includes those mid-career professionals returning to analytical work after years focused on policy or management, recent graduates trying to deepen their skills on the job, and even technical staff seeking more structure in how they approach environmental data analysis.

Climate change is one of the most complex and consequential challenges humanity faces. It is not a problem that can be solved solely with data or models—but neither can it be addressed without them. As practitioners, we must bring together the best available knowledge, the most reliable tools, and the clearest thinking we can muster.

My hope is that this book helps demystify the analytical process and empowers professionals to use data more confidently and effectively in their work.

In short, this is a book written by a professor who has spent a career straddling the worlds of theory and practice—and who believes deeply in the value of equipping practitioners with the tools they need to make sense of a changing world. I invite you to use this handbook not as a strict manual, but as a companion: one that you can refer to, learn from, and adapt as you navigate your own path through the evolving landscape of climate adaptation analytics.

Let's get to work.

Monterey, California, USA Fernando DePaolis

Acknowledgements I want to recognize all the help and support my wife provided, not just when I was writing this book, but always.

I am grateful to my Middlebury Institute colleagues, who encouraged me throughout the process of writing this book. A special recognition goes to my friend and colleague, Prof. Robert McCleery, for his assistance in reviewing the full manuscript. Of course, all remaining mistakes are my own.

Introduction: Getting Started with R and RStudio

R, What it is, and its History

R is a programming language and environment for statistical computing and graphics, developed in the early 1990s by statisticians Ross Ihaka and Robert Gentleman at the University of Auckland, New Zealand. Designed as a free and open-source alternative to proprietary statistical software (such as SPSS, Stata and many others), R rapidly gained traction in academic and research communities due to its extensibility, rich package ecosystem, and active user base. Over the years, R has evolved into a powerful tool for data analysis, modeling, and visualization—qualities that make it especially valuable in addressing complex, interdisciplinary challenges like climate change adaptation.

The history of R is one of those fascinating stories in the world of computer software. We can trace its origin to another language, "S." John Chambers and others, at the time working at the Bell Telephone Labs (part of AT&T Corp), started the development of this environment almost 50 years ago, in 1976, as an internal project for statistical analysis.

Roger Peng [1] has a comprehensive discussion of all the multiple twists and turns that S underwent and the origin of R. In my opinion, the most significant event was the transformation of R into a platform that embraces the contemporary definition of 'open source.' Open source means that the software is not only free of charge but also—and most importantly—anyone can tinker with its code. Such a momentous shift happened in 1995 when Martin Mächler persuaded the developers of R (Ihaka and Gentleman) to adopt the GNU General Public License. Since then, R has grown explosively into a vast environment with thousands of packages built by the topmost experts in a wide variety of fields.

Peng's discussion of R's limitations is both effective and comprehensive. It is important to understand those areas in which R does not perform well, such as memory management (especially for large datasets); resource allocation (it can be difficult to take full advantage of multi-core systems); integration into production environments; and security issues (it is open source, after all). If these issues are a concern for your work environment, I strongly suggest you seek out support from your IT colleagues.

Additionally, for those of you interested in a comprehensive list of analytical re-sources, I recommend the Comprehensive R Archive Network (CRAN) project website (https://cran.r-project.org/). A resource that can be particularly useful when familiarizing with R is the *CRAN Task Views*. These are aggregations of packages into coherent groups by scientific field or area of professional practice. You can visit it at https://cran.r-project.org/web/views/.

RStudio

RStudio, a flagship product of Posit (formerly RStudio Inc.), was established in 2009 to create a more intuitive and integrated development environment (IDE) for users of the R programming language. It has since become the de-facto standard IDE for R practitioners worldwide, revolutionizing the way data scientists work and collaborate across diverse applied fields, as well as basic science environments.

An IDE is a software application that aggregates various tools essential for software development. While many IDEs exist for different programming languages—and some are even language-agnostic—RStudio stands out for its deep integration with R. Its comprehensive feature set includes code highlighting, seamless version control integration, and robust support for reproducible research through powerful tools such as R Markdown and Shiny. This focus on reproducibility, combined with an open-source philosophy and a strong emphasis on collaboration, has significantly enhanced the capabilities of data science. RStudio also provides integration with package development applications, as well as versioning via GitHub.

Beyond the IDE, Posit is a key contributor to the R ecosystem, developing and maintaining many of the most widely used R packages. These include foundational packages like `tidyverse` and `ggplot2`, specialized libraries such as `tidymodels` for machine learning, and interfaces like TensorFlow for R, among many others.

Posit supports both experienced and novice users of R and RStudio. They provide an extensive array of resources, with the RStudio User Guide (https://docs.posit.co/ide/user/) being particularly valuable. This document guides the data science practitioner on setting up the environment, optimizing productivity, and developing their own R packages, empowering users to become more effective and efficient R developers.

Getting Started with This Book

The most effective way to follow this book is to install R and RStudio on your computer. The installation is straightforward, and both pieces of software run on any contemporary operating system (Windows, macOS, and any implementation of Linux).

In the Appendix, you will find a list of the necessary libraries and a recommended code block to install and load them. It is possible, depending on how your computer and software are set up, that you will need to install some other libraries. Also, you may find minor discrepancies between the code blocks in this book and the ones you need to run on your computer, especially if you are running R on Windows.

You can follow the book from its companion website, where you can find all the code blocks and data files.

Reference

1. Peng, R. 2016. *R Programming for Data Science*. ISBN: 1365056821. https://bookdown.org/rdpeng/rprogdatascience/.

Contents

About the Author

Dr. Fernando DePaolis is a professor at the Middlebury Institute of International Studies in Monterey, California, where he teaches advanced graduate courses in environmental analysis and management. He holds a doctorate in Urban Planning from UCLA and brings extensive research experience to his academic work.

Dr. DePaolis has served as a consultant for government agencies and organizations at the local, regional, and national levels, providing expertise in environmental planning and policy analysis. His specialization in quantitative analysis from an applied perspective bridges the gap between academic theory and professional practice, offering students and colleagues valuable insights into real-world applications across diverse fields.

Through his combined experience in academia, research, and consulting, Dr. DePaolis contributes a unique practitioner's perspective to the study of environmental management and urban planning challenges.

List of Figures

List of Tables

Part I
Describing and Visualizing Data

Chapter 1
Descriptive Statistics

1.1 Introduction

In any data-centric workflow, and climate adaptation analytics is not an exception, it is crucial to begin by thoroughly understanding the underlying data. Descriptive statistics provide a foundational layer for this understanding by summarizing and revealing the main features and structure of a dataset. This chapter introduces the key descriptive statistics and basic data exploration techniques used to assess data quality and suitability for further analysis in R. These methods form a bridge between data acquisition and the application of advanced statistical or geospatial models, ensuring analytical robustness and interpretive clarity.

1.2 Data Types and Structures

Data types

We want to introduce a basic typology of data, structured around two principal categories: *metric* and *categorical* data. These categories, often referred to as "levels of measurement," are fundamental to understanding the nature of empirical observations. Metric data, characterized by their quantitative properties, are further delineated into *ratio* and *interval* levels. Ratio data possess a true zero point, allowing for meaningful ratios (e.g., height, weight), while interval data lack a true zero, though they maintain consistent intervals between values (e.g., temperature in Celsius or Fahrenheit).

Conversely, categorical data, which represent qualitative distinctions, are subdivided into *ordinal* and *nominal* levels. Ordinal data exhibit a natural order or ranking among categories (e.g., educational levels: high school, bachelor's, master's; preferences: agree, neutral, disagree), but the intervals between categories are not necessarily uniform or meaningful. Nominal data, the most basic form of categorical data, consist of categories without any inherent order or ranking (e.g., gender, marital status, place of residence, etc.).

F. DePaolis, *Workflows for Climate Change Adaptation Analytics*,
https://doi.org/10.1007/978-3-032-14004-3_1

We can also refer to the amount of information embedded in each type. Thus, ratio variables would contain–and communicate–the highest amount of information, while nominal data would contain the lowest amount. There is also the issue of convertibility. We can always convert metric variables into ordinal or nominal, but the opposite is not true: we cannot reconstruct the full scale of a metric variable from information captured in nominal variables.

Our intention is not to provide an exhaustive analysis of data types, levels of measurement or information theory. However, the true significance of these data types extends far beyond mere categorization. Understanding the specific level of measurement for your variables is absolutely critical for the selection of appropriate statistical methods. The choice of statistical analysis techniques—-whether descriptive or inferential—is directly dictated by the properties inherent in each data type. Employing an unsuitable statistical method for a given data type can lead to erroneous conclusions, misinterpretations of findings, and ultimately, undermine the rigor and validity of any empirical analysis. Therefore, a deep appreciation of these data types and their implications is not merely academic; it is an indispensable prerequisite for conducting sound and insightful quantitative research.

Data structures in R

Regardless of the data type, R has the ability to handle a myriad of data objects and structures. Let's take a look at some of them. In R, vectors, lists, matrices, arrays, and data frames are the fundamental and most important data structures. Vectors are the simplest, holding a sequence of elements of the *same* type. Lists can hold *different* types of elements. Matrices are two-dimensional arrays of the same type, and data frames are tabular structures commonly used for storing data of different types. For additional information, see Grosser, Bumann, and Wickham [1]. Here's a more detailed breakdown:

- **Vectors**: The most basic data structure in R, holding a sequence of elements of the same type (numeric, character, logical, etc.).
- **Lists**: A more flexible data structure that can contain different types of objects, including vectors, matrices, and other lists.
- **Matrices**: Two-dimensional arrays, where all elements must be of the same type.
- **Arrays**: Multi-dimensional arrays, extending the concept of matrices to more than two dimensions.
- **Data Frames**: Tabular data structures, commonly used for storing datasets. They consist of columns (variables) and rows (observations, defined by the unit of analysis) and can be created from data files or using the `data.frame()` function.

These data structures are essential for working in R, providing ways to organize, manipulate, and analyze information effectively. In addition to these basic structures, R comprises a large number of objects containing both data and the results of data processing and statistical tests. The number of objects is unlimited and keeps growing as developers build new functions and objects. Throughout the book, we make use–and add brief explanations–of some of those objects.

1.3 Understanding Datasets

In the late 1970s, the famous statistician Tukey [2] coined the term *Exploratory Data Analysis* (EDA) to refer to the process of investigating datasets to discover patterns, spot anomalies, test hypotheses, and check assumptions through statistical summaries and graphical representations. More than an analytical technique, EDA is a mindset in which the analyst is always on the lookout for less-than-obvious features in the data that could be the gateway to a significant breakthrough.

Before proceeding to complex modeling or spatial visualization, it is essential to conduct a structured assessment of the dataset. This includes confirming the data's origin, structure, and quality. If the dataset was not collected by the analyst, best practices recommend obtaining metadata—a formal description of the dataset's characteristics. Metadata typically includes:

- A data dictionary explaining each variable's name, measurement units, and data types.
- The unit of analysis (e.g., individual, household, county).
- The time frame of data collection and update schedules.
- Details on data provenance, including who collected the data and for what purpose.
- Any known data quality issues or limitations (e.g., sampling errors, sample size, etc.).

High-quality metadata ensures transparency and reproducibility and is a key requirement in climate and environmental sciences, where datasets are often compiled from remote sensing, administrative sources, or sensor networks [3].

In addition to understanding metadata, it is important to examine the data structure directly. When working with tabular data, one should confirm how variables relate to units of observation. For instance, in census or economic data, each row might represent a county-year combination, with attributes such as population or the value of economic output. Analysts must assess data completeness, identify missing values, and recognize whether the dataset is dense (few missing values) or sparse.

A further distinction must be made between *wide* and *long* data formats [4]:

- In **wide** format, each unit of observation occupies one row with separate columns for each time or category (e.g., yearly temperatures in separate columns).
- In **long** format, multiple rows may correspond to the same unit, with an additional variable denoting time or category.

Many R functions and packages (such as `ggplot2` and `tidyverse`) expect data in long format for efficient manipulation and visualization. Transforming between these structures is often necessary using functions like `pivot_longer()` or `pivot_wider()` from the `tidyr` package, or `mutate()` and other functions in the `dplyr` package.

Finally, before applying statistical techniques, one must confirm that the dataset meets the assumptions required by those methods—such as normality, independence, sampling schema, or minimum sample size. Ignoring these requirements can compromise model validity and lead to misleading conclusions [5].

1.4 Summary Measures

Introductory statistics curricula often begin with descriptive statistics, which summarize the key attributes of variables without making inferential claims. These include:

- Measures of Central Tendency, such as:
  - Mean ($\bar{x}$): arithmetic average
  - Median: the middle value when data is sorted
  - Mode: the most frequently occurring value
- Measures of Dispersion, such as:
  - Range: difference between maximum and minimum
  - Variance (σ^2): average squared deviation from the mean
  - Standard Deviation (σ): square root of variance
  - Interquartile Range (IQR): difference between 75th and 25th percentiles

These statistics help identify patterns, detect anomalies, and provide baselines for modeling. In the R environment, most of these measures can be calculated using base functions such as `mean()`, `sd()`, and `summary()`. Additional functionality is available through packages such as `psych`, `skimr`, and `Hmisc`, which offer enhanced summaries and nicely formatted outputs suitable for reporting.

Some packages—such as `dplyr`, `data.table`, and `janitor`—also provide convenient data manipulation capabilities that complement statistical analysis. The `tidyverse` collection, in particular, is highly suited to chaining operations using the pipe operator (|> or %>%) for clear and more efficient workflows. By *piping*, one can skip the creation of separate data objects that will not be necessary later on.

Descriptive statistics can also be embedded in visualizations, which are explored further in the next section. Graphical summaries such as boxplots, histograms, and density plots provide immediate insight into data distributions, helping validate the numeric summaries and expose distributional issues (e.g., skewness, multimodality).

Examples

To illustrate the use of descriptive statistics in a climate-economic context, we consider data from the National Ocean Economics Program—NOEP— [6]. The NOEP provides structured datasets describing ocean-related economic activity along U.S. coastal regions and the Great Lakes. These datasets include variables on employment, establishments, and wages, categorized by ocean sectors (e.g., tourism, transportation, marine construction).

These are the key variables that we use from the NOEP dataset:

- `geo_id`: Geographic FIPS code[1]
- `state`, `county`: Location identifiers
- `sector`, `industry`: Economic categories
- `employment`, `establishments`, `wages`: Economic indicators

---

[1] FIPS, or Federal Information Processing Standards, are standards and guidelines developed by the National Institute of Standards and Technology (NIST) for multiple applications. There is a distinct FIPS code for each county in the United States.

We preprocess the dataset by removing irrelevant variables (e.g., `source_gdp`, `rgdp`, `flag`) and filtering out entries with zero establishments. Additionally, we removed the "Ocean Economy" sector, as it duplicates information from other sectors by aggregating sectoral totals.

We compute summary statistics across all U.S. counties, and then extract a subset corresponding to California counties. Using grouped summaries, we calculate the mean, standard deviation, and count for average wage by sector, a variable we will create later. This approach reveals both central trends and inter-sectoral variability.

Notably, while California has 58 counties, not all counties are represented in each ocean sector. In total, the dataset contains 90 county-sector observations spanning 23 unique counties. Such disaggregation is essential for region-specific climate adaptation planning, where economic vulnerability and exposure differ by sector and geography.

First, we need to read the data from the NOEP site using an API (Application Programming Interface)[2] to access it. The following code block accesses the website and collects the data in a dataframe.

```
NOEP_2021 = 'https://oceaneconomics.org/Api/Ocean/econData?years=2021'
OceanEcon <- GET(NOEP_2021)
OceanEcon_US = fromJSON(rawToChar(OceanEcon$content))
```

JavaScript Object Notation (JSON) is a lightweight, text-based format for data interchange that has become a cornerstone of modern computing, particularly in web development and data-driven applications. Designed for both human readability and machine parsing, JSON facilitates seamless communication between systems, especially between web clients and servers [7]. One of its primary advantages lies in its language-agnostic design, allowing data to be easily exchanged between heterogeneous systems, such as those written in Python, JavaScript, or Java. JSON has become ubiquitous in web services and APIs, where it enables structured, hierarchical data to be transmitted efficiently over the Internet. There are many resources available online with details about JSON. We encourage the reader to explore them, as the usage of this format is ubiquitous.

Here, we list the variable names and types.

```
var_names <- data.frame(
  variable = names(OceanEcon_US),
  class = sapply(OceanEcon_US, class)
)
```

The code below creates Table 1.1.

```
kable(var_names, format = "latex", booktabs = TRUE,
     linesep = "", row.names = FALSE,
     caption = "List of variables in the Ocean Economy dataset") |>
  kable_styling(latex_options = c("striped", "HOLD_position", "scale_down"))
```

We can see that the dataset has variables for geographic descriptors (`geo_id` is the county FIPS code, `state`, and `county`), industry descriptors (`sectors` and `industry`), and economic indicators (`establishments`, `employment`, `wages`, etc.) We refer the reader to the NOEP website for additional details.

[2] An API is a set of rules and protocols that defines how different software components or applications can communicate and interact with each other.

Table 1.1 List of variables in the ocean economy dataset

Variable	Class
geo_id	Character
Year	Integer
State	Character
County	Character
Source	Character
Sector	Character
Industry	Character
Establishments	Integer
Employment	Integer
Wages	Numeric
gdp	Numeric
rgdp	Numeric
flag	Integer

In the next step, we clean up the dataset we just downloaded. We will remove the variables `source` `gdp`, `rgdp`, and `flag`, as we do not need them at this time. The code also removes the "Ocean Economy" sector.

```
OceanEcon_US <- OceanEcon_US |>
  select(1:4, 6:10) |>
  dplyr::filter(
    substr(geo_id, 3, 5) != "000",
    establishments != 0,
    sector != "Ocean Economy"
  )
```

We will now create two new variables (`avrg_wage` and `est_size`) computed from existing variables.

`avrg_wage`: Average wage per employee as:

$$avrg\_wage = \frac{wages}{employment}$$

`est_size`: Average establishment size (employment per establishment) as:

$$est\_size = \frac{employment}{establishments}$$

```
# Average wage per worker
OceanEcon_US$avrg_wage <- OceanEcon_US$wages / OceanEcon_US$employment

# Average employment per establishment
OceanEcon_US$est_size <- OceanEcon_US$employment / OceanEcon_US$establishments
```

We are now ready to get some descriptive statistics for the variables in the dataset for all the counties in the United States. Tables 1.2 and 1.3 show the results of the `sumtable` and `summary` functions, respectively.

```
sumtable(OceanEcon_US[,c(8,10,11)],
   title = "Summary Statistics using 'sumtable' [United States]",
   out= "kable")
```

The code below creates a table with the summary statistics for the United States' ocean economy.

```
table2 <- summary(OceanEcon_US[,c(8,10,11)])
```

Table 1.2 Summary statistics using 'sumtable' [United States]

Variable	N	Mean	Std. Dev.	Min	Pctl. 25	Pctl. 75	Max
Employment	1011	2901	8042	0	77	1806	124315
avrg_wage	1003	51799	32386	11344	27627	66803	295248
est_size	1011	23	63	0	5.5	19	1082

Table 1.3 Summary statistics using 'summary' [United States]

Employment	avrg_wage	est_size
Min. : 0	Min. : 11344	Min. : 0.00
1st Qu.: 77	1st Qu.: 27627	1st Qu.: 5.46
Median : 330	Median : 45469	Median : 11.43
Mean : 2901	Mean : 51799	Mean : 22.68
3rd Qu.: 1806	3rd Qu.: 66803	3rd Qu.: 18.65
Max. :124315	Max. :295248	Max. :1082.00
NA	NA's :8	NA

Table 1.4 Summary Statistics using 'sumtable' [California]

Variable	N	Mean	Std. Dev.	Min	Pctl. 25	Pctl. 75	Max
Employment	90	5469	12738	5	102	4343	80198
avrg_wage	90	70053	36405	25179	40382	93424	205326
est_size	90	22	35	0.39	6.3	19	254

```
kable(table2, digits=2,
    caption = "Summary Statistics using 'summary' [United States]",
    format="latex", booktabs=T, linesep="") |>
  kable_styling(latex_options = c("striped", "HOLD_position", "scale_down"))
```

Notice that the code above includes an option `format="latex"`. This is the code that generates the typesetting for the printed version of the book. If the `format` clause is omitted, the table will show on screen.

The dataset for California is a subset of the dataset for the United States. The following code block extracts the data for California and generates the descriptive statistics. Table 1.4 displays the summary statistics for California, using the function `sumtable`.

```
OceanEcon_CA <- OceanEcon_US |>
  dplyr::filter(state == "California") # Notice that we use two equal signs
```

```
sumtable(OceanEcon_CA[,c(8,10,11)],
   title = "Summary Statistics using 'sumtable' [California]",
   out = "kable")
```

We now introduce a modification to the listing of summary statistics. We will aggregate the California data by sectors of the Ocean Economy, and report the mean, standard deviation, and number of observations for a single variable, *average wage per employee*.

Table 1.5 Summary statistics for 'Average Wage' by Sector [California]

Sector	Mean	Std deviation	N
Living Resources	57654.78	19204.45	19
Marine Construction	102210.84	35859.68	18
Marine Transportation	68638.59	23177.81	22
Offshore Mineral Extraction	114869.76	36155.65	9
Ship and Boat Building	71610.70	11966.98	3
Tourism and Recreation	32149.71	5094.03	19

Regarding the number of observations, there is an important aspect to consider. The value of n reported in the tables corresponds to combinations of counties and sectors. In total, there are six sectors in the Ocean Economy, and 58 counties in the state of California (for a possible total of 348 county-sector combinations). However, not all sectors are present in all counties, and not all counties are included in the Ocean Economy, as some of them do not qualify as coastal counties. There are 23 unique counties in the dataset, for a total of 90 county-sector combinations.

```
table_results_CA <- OceanEcon_CA |>
  group_by(sector) |>
  summarise(Average_wage = mean(avrg_wage, na.rm = TRUE),
            StDev_wage = sd(avrg_wage, na.rm = TRUE),
            Observations = n(),
            .groups = "drop"
  )
```

We can format the output using the parameters in the `kable` function to create the final version of the table with the descriptive statistics for Average Wages by Sector, as shown in Table 1.5.

Please notice that developers and programmers can always further optimize chunks of code (in R or any other language) to take advantage of new packages or design approaches. For example, ubiquitous platforms like *GitHub Copilot* now support co-creation by suggesting snippet completion. In the near future, they will not just provide options; they will understand the broader goals, design patterns, and even business logic. We invite the reader to experiment with alternative formulations of the code chunks in this book by using AI-based systems.

```
kable(table_results_CA, digits=2,
  caption = "Summary Statistics for 'Average Wage' by Sector [California]",
  col.names = c("Sector", "Mean", "Std deviation", "N"),
  format="latex", booktabs=T, linesep="") |>
  kable_styling(latex_options = c("striped", "HOLD_position", "scale_down"))
```

References

1. Grosser, Malte, Henning Bumann, and Hadley Wickham. 2022. *Advanced R Solutions*. The R Series. Boca Raton, FL: Chapman and Hall/CRC.
2. Tukey, John W. 1977. *Exploratory Data Analysis*. Reading, MA: Addison-Wesley Publishing Company, pp. xvi.

3. National Research Council. 2007. *Environmental Data Management at NOAA: Archiving, Stewardship, and Access*. Washington, DC: The National Academies Press. https://doi.org/10.17226/12017. https://nap.nationalacademies.org/catalog/12017/environmental-data-management-at-noaa-archiving-stewardship-and-access. ISBN: 978-0-309-11209-3.
4. Wickham, Hadley. 2014. Tidy Data. *Journal of Statistical Software* 59(10): 1–23. https://doi.org/10.18637/jss.v059.i10.
5. Wilks, Daniel S. 2011. *Statistical Methods in the Atmospheric Sciences*. Amsterdam; Boston: Elsevier Academic Press. ISBN: 9780123850225 0123850223.
6. Center for the Blue Economy. 2012. *National Ocean Economics Program*. https://www.oceaneconomics.org/.
7. Bray, Tim. 2014. *The JavaScript Object Notation (JSON) Data Interchange Format*. RFC 7159. https://doi.org/10.17487/RFC7159. https://www.rfc-editor.org/info/rfc7159.

Chapter 2
Visualization

2.1 Introduction

As computers advanced in their ability to produce high-quality graphics, the 1980s saw a distinct bibliographical trend aimed at helping both developers and the general public understand how we interpret visual information. Tufte [1] emerged as a leading figure, with his best-selling books significantly shaping our appreciation of the power of graphic communication. This body of work has continued to grow through contributions from others, both in the early years and more recently (Cleveland [2] and Few [3]), in conjunction with the evolution of hardware and software.

For all the computer power and modeling capabilities we have today, it is still true that a picture is worth a thousand words. In this section, we will use graphics to understand the distribution of values in our datasets. Since its inception, the R environment has emphasized graphic representation as a significant measure of data analytics. For example, the graphing package `ggplot2`, developed by Wickham [4], is so powerful that it counts as a stand-alone data analysis platform. We encourage the reader to consult the multiple resources available on it and similar packages.

The psychological and physiological mechanisms by which we perceive and assign meaning to images have been the subject of study for a long time. Cleveland and McGill [5] provide a series of coherent conjectures (not quite a theory by their own account) that organize those perceptual processes. Academic fields and professional practice as diverse as advertising, landscape design, and education use these foundations for all kinds of purposes, from noble to deceptive. Quantitative analytics is not that different in the sense that we could use clever techniques to either enhance our audience's understanding of a complex phenomenon or, conversely, unduly influence their perception in a particular direction.

In this section, we complement the numeric descriptive statistics we computed in the previous section with some of the most commonly used graphics. Although there is a wide variety of visualization options, here we focus on the contribution that graphics make as part of descriptive analytics. We continue using the NOEP dataset

F. DePaolis, *Workflows for Climate Change Adaptation Analytics*,
https://doi.org/10.1007/978-3-032-14004-3_2

and build graphic representations for some of the variables in the United States and California Ocean Economy datasets.

A note on the relationship between data types and graphic options

First, let me add a few comments on the most appropriate visualization approaches and their relationship to the different types of data we discussed earlier.

We already referred to two main types of data: metric and categorical. The defining characteristic of **metric** data is its *continuity*—the mathematical property that infinite possible values exist between any two measured points. For instance, between temperatures of 20 and 21 °C, we can theoretically measure 20.1, 20.15, 20.153 °C, and so forth, limited only by the precision of our instruments. This continuity reflects the underlying nature of the phenomenon being measured, whether it's temperature, time, weight, or distance.

This continuous nature has profound implications for visualization. When we display metric data, we are representing not just discrete data points but the continuous relationship between values across the entire range of measurement. This attribute makes the *line chart*—in all its complex variations—the preeminent chart type to use with metric data. We will use them in the chapter on time series to represent a variable's change over time. Variables such as temperature variations throughout the day and housing stock across decades are just a few examples.

Conversely, **categorical** data, where the spaces between categories are meaningful separations rather than continuous transitions, requires a different type of graphics. Categorical data represent distinct groups, classifications, or nominal categories that cannot be mathematically interpolated. To represent them, we predominantly use *bar charts*, regardless of the specific graphic details we employ.

Finally, we refer to the ubiquitous but often misused *pie charts*. They occupy a unique position in data visualization and should be reserved for very specific circumstances. They are appropriate only when displaying data that represents parts of a meaningful whole, proportional relationships, or the presence of mutually exclusive categories.

This is by no means an exhaustive discussion of the vast field of graphic representation and data visualization, but it will serve, we hope, as a starting point for the reader's own exploration of such a fascinating topic.

2.2 Data

Again, we use the API for the NOEP database, but this time we use an optimized version of the data acquisition code. This way, we can get the data in the format we need by removing unnecessary variables and sectors, and creating the new ones we need, all in a single block of code. We then create a dataset for California by filtering the United States dataset.

```
# Download and parse data
NOEP_2021 <- 'https://oceaneconomics.org/Api/Ocean/econData?years=2021'
OceanEcon_US <- GET(NOEP_2021) |>
  content("text") |>
  fromJSON() |>
  as.data.frame() |>
  select(1:4, 6:10) |>
  filter(
    substring(geo_id, 3, 5) != "000",
    establishments != 0,
    sector != "Ocean Economy"
  ) |>
  mutate(
    avrg_wage = wages / employment,
    est_size = employment / establishments
  )

# Subset for California
OceanEcon_CA <- filter(OceanEcon_US, state == "California")
```

2.3 Colors

Let us first set up some colors that we will use in the graphics below. There are multiple ways to specify the colors in R. The simplest one is to refer to them by their natural language names, such as *blue* or *green*. A variety of institutions and research groups keep online resources with lists of color names you can use in R. You can also use the standard hexadecimal codes corresponding to each color. The *hex* codes, as they are commonly known, are fully equivalent to the ones in the RGB and HSL systems[1] and are preferred in the HTML coding environment. The following code assigns names to six different hex codes for shades of blue, orange, brown, red, purple, and gray. We were inspired by the colors traditionally associated with the native cultures of the Southwest of the United States.

```
SWblue <- "#719f9f"
SWorange <- "#ed8b22"
SWbrown <- "#b58c44"
SWred <- "#CD5C5C"
SWpurple <- "#804338"
SWgray <- "#747474"

SoWest <- c(SWblue, SWorange, SWbrown, SWred, SWpurple, SWgray)
```

[1] Color systems like *RGB* (Red, Green, Blue) and *HSL* (Hue, Saturation, Lightness) are ways to describe colors mathematically.

2.4 Boxplots

A **boxplot** (also known as a *box & whisker* plot) is a graphic representation of some of the basic descriptive statistics for a single variable. In general, they show the minimum and maximum values, the mean, the 25th and 75th percentiles (also known as 1st and 3rd quartiles, or simply Q1 and Q3), as well as any outliers. We show one in Fig. 2.1. For each of the ocean economy sectors, Fig. 2.1 shows the median (the solid line inside the box), the 25th and 75th percentiles, and the outliers. The thin, dotted lines—sometimes called *whiskers*—represent the theoretical minimum and maximum computed as the $Q1 * IQR$ and $Q3 * IQR$. *IQR* stands for inter-quartile range.

There are alternative ways of generating a boxplot. The first one (in Fig. 2.2) is generated by the function `boxplot` in the package `graphics`, one of the basic components of R.

```
boxplot(avrg_wage ~ sector, data = OceanEcon_US, lwd = 0.5,
      boxlwd = 0.75, xaxt="n",
      main="Average Wage by Sector in the United States Ocean Economy",
      xlab="Sector", ylab="Avrg Wage", col = SoWest,
     cex.lab=0.6, cex.main=0.7, cex.axis=0.6)
axis(1,at=1:6,labels=F)
nnames<-c("Living Resources","Marine Construction","Marine Transportation",
       "Offshore Mineral Extraction", "Ship and Boat Building",
       "Tourism and Recreation")
mtext(nnames,1,cex=0.5,at=1:6,line=rep(c(1,2),6))
```

Another way is to use the functions in the `ggplot2` package, displayed in Fig. 2.3. In this case, we use the parameter `coord_flip()` to plot the boxes horizontally,

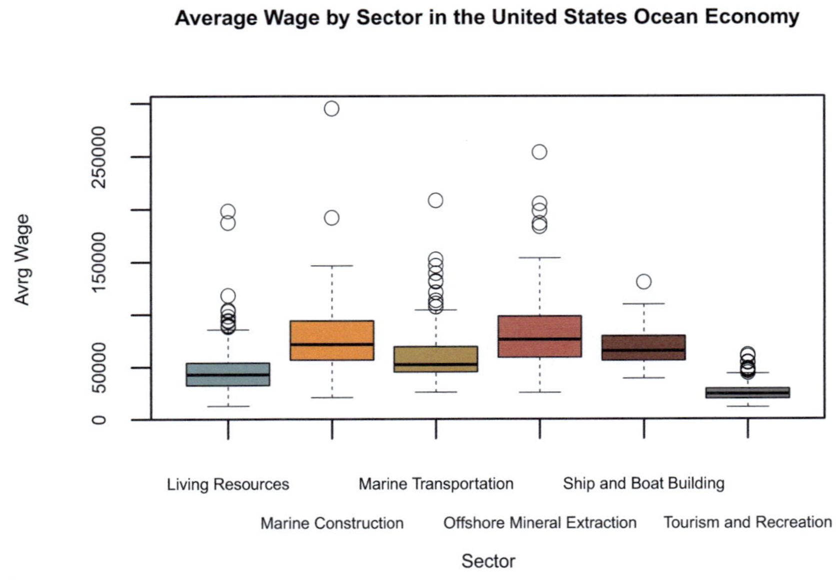

Fig. 2.1 Boxplots using the function in the graphics package (US)

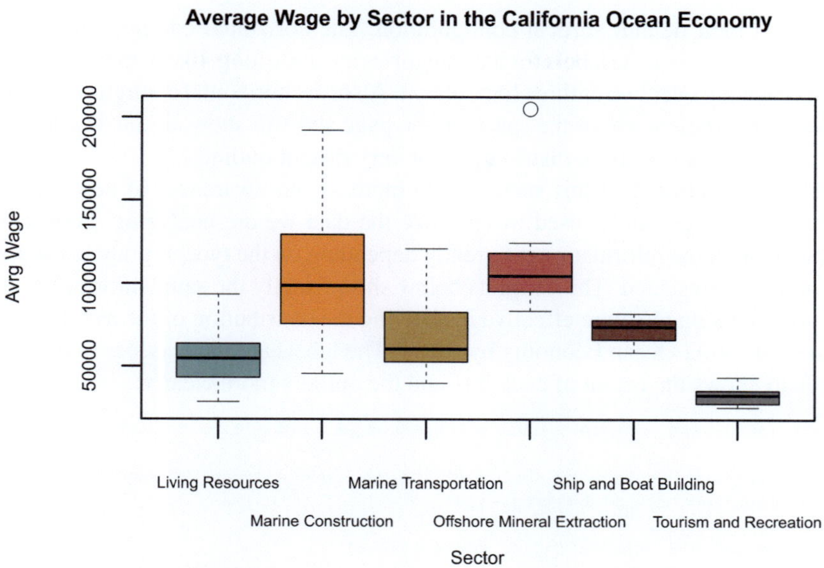

Fig. 2.2 Boxplot using the function in the graphics package (California)

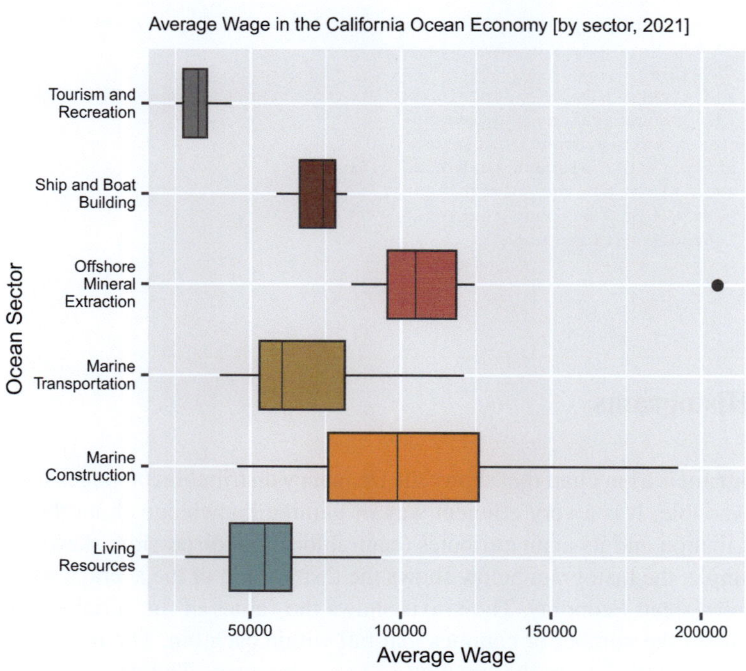

Fig. 2.3 Boxplot using the ggplot package horizontal setup

instead of their default vertical configuration. The horizontal configuration can be quite effective when the labels for the categories are quite long, like in this case, where the labels wrap over two or three lines of text. Also, the horizontal configuration could make more efficient use of the space on the page and still show all the details of the boxplot, such as very tight distributions or very distant outliers.

The main purpose of this section is to increase our awareness of how different types of graphics can be used to visualize the data we are analyzing. People will perceive the same information differently depending on the type of graphic used and the way it is presented. The boxplots below show exactly the same information, but the horizontal one is more effective in showing the distribution of the average wage in the California Ocean Economy by sector. The labels are much easier to read, and the chart shows the extent of each box and the outliers more clearly.

```
boxplot(avrg_wage ~ sector, data = OceanEcon_CA, lwd = 0.5, boxlwd = 0.75,
    xaxt="n",
      main="Average Wage by Sector in the California Ocean Economy",
      xlab="Sector", ylab="Avrg Wage",
      col = SoWest,
      cex.lab=0.6, cex.main=0.7, cex.axis=0.6)
axis(1,at=1:6,labels=F)
mtext(nnames,1,cex=0.5, at=1:6,line=rep(c(1,2),5))
```

```
ggplot(OceanEcon_CA, aes(x = sector, y = avrg_wage, fill=sector)) +
  geom_boxplot(lwd=0.4, fatten = 0.6) +
  labs(title="Average Wage in the California Ocean Economy [by sector, 2021]",
      x="Ocean Sector", y = "Average Wage") +
  coord_flip() +
  scale_x_discrete(labels = function(x) stringr::str_wrap(x, width = 14))+
  scale_fill_manual(values = SoWest) +
    theme(axis.text=element_text(size=6),
          strip.text = element_text(size = 6),
          axis.title=element_text(size=8),
          plot.title = element_text(size = 7),
          legend.position="none"
    )
```

2.5 Histograms

A **histogram** is a bar chart that shows the frequency distribution of values for a single metric variable. It is a very efficient way of gaining knowledge about the shape of the distribution and its main attributes (central location, dispersion, skewness, etc.). For example, the histogram below shows the distribution of the average wage in the California Ocean Economy. The x-axis shows the values of the variable, while the y-axis shows the number of counties that fall within each bin. The bins are defined by the parameter `bins` in the function `geom_histogram`. The default value is 30, but we can change it to any other value, as shown below, where we use the `bins=20` option.

```
ggplot(OceanEcon_CA, aes(x = avrg_wage)) + ylim(0,15)+
  geom_histogram(size = 0.2,colour=SWgray,fill=SWblue,alpha=0.6, bins = 20)+
  labs(title="Average Wage in the California Ocean Economy [totals, 2021]",
     x="Wage [dollars]", y = "Number of Counties")+
   theme(axis.text=element_text(size=6),
        strip.text = element_text(size = 7),
        axis.title=element_text(size=7),
        plot.title = element_text(size = 8)
   )
```

The histogram of the *Average wage per employee* in the Ocean Economy for the state of California in Fig. 2.4 displays positive (or right-sided) skewness. That is, the bulk of values are concentrated in the lower end of the wage spectrum, with the majority of cases located below the $100,000/year mark. This is a typical distribution of wages, as the predominant majority of workers earn lower wages and relatively few earn higher wages. In that regard, we can conclude that the distribution displays the expected behavior.

We can also create a **panel** of histograms for the same variable, grouped by a factor (categorical variable), as the one in Fig. 2.5. We used the function `facet_wrap` to create the panel of histograms, one for each sector in the California Ocean Economy. The parameter `facet_wrap` is used to specify the variable used to create the panels. In this case, we use the variable `sector`, which is a categorical variable with six different values. The function `facet_wrap` will create a separate histogram for each

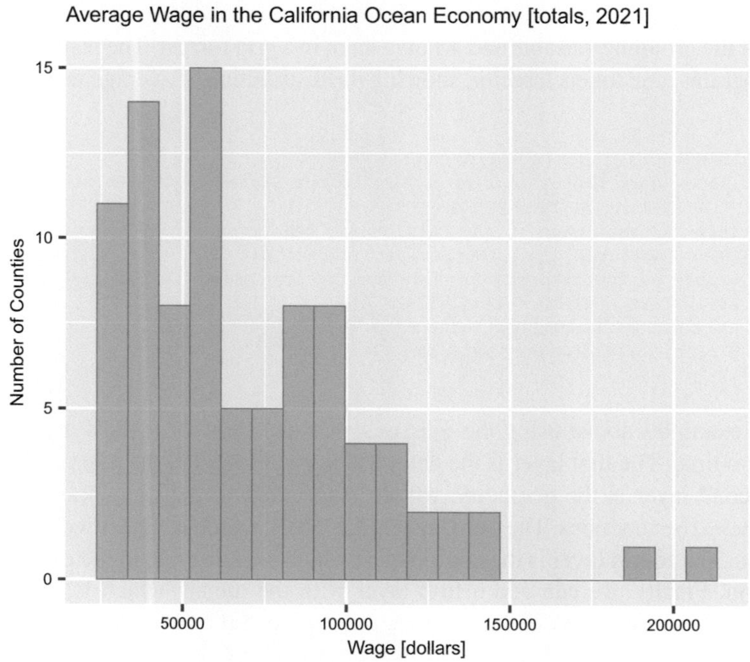

Fig. 2.4 Histogram of a single variable (average wage)

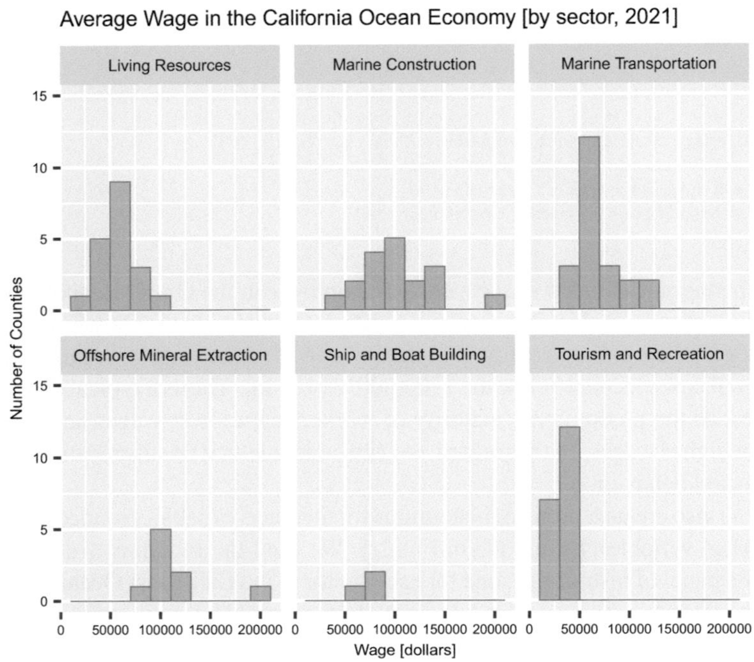

Fig. 2.5 A panel of histograms

value of the grouping variable and arrange them in a grid format. The result is a panel of histograms, one for each sector, showing the distribution of average wages in each sector.

```
ggplot(OceanEcon_CA, aes(x = avrg_wage)) +
  ylim(0,15) + geom_histogram(size = 0.2, colour=SWgray, fill=SWblue,
              alpha=0.6, bins = 10) +
  labs(title="Average Wage in the California Ocean Economy [by sector, 2021]",
      x="Wage [dollars]", y = "Number of Counties")+
  facet_wrap(~ as.factor(sector))+theme(axis.text=element_text(size=5),
        strip.text = element_text(size = 6),
        axis.title=element_text(size=6),
        plot.title = element_text(size = 8)
        )
```

The examples above using the `ggplot2` package show how plots are built one layer at a time. The first layer is the data, which is specified in the `ggplot` function. The second layer is the geometry, which is specified in the `geom_histogram` or `geom_boxplot` functions. The third layer is the labels, which are specified in the `labs` function. The fourth layer is the coordinates, which are specified in the `coord_flip` function. Finally, we can add a fifth layer with the theme, which is specified in the `theme` function. The `ggplot2` package is a powerful and flexible tool to create graphics in R, allowing us to customize every aspect of the plot. The *gg* part of `ggplot` stands for "grammar of graphics", which is a way to describe the components of a graphic in a systematic way [4].

We can also use the `ggplot2` package to create all kinds of graphics and plots, such as scatter plots, line plots, bar charts, and more. The package is very flexible and allows us to create complex graphics by combining simple components. For additional content and examples, we recommend Wickham [4]. This book is a great resource for learning how to use R for data analysis and visualization, and it covers many of the topics we have discussed in this section in more detail.

2.6 Density Plots

Finally, we create *density plots* of the same variable. Density plots or *density curves* are smoothed versions of histograms. We are including them not because they can better convey the same information as a histogram, but as an example of how easy it is to get distracted by fancier and fancier versions of the same charts.

First, in Fig. 2.6, we plot all the curves in a single chart, and then, in Fig. 2.7, we plot them in a grid, in a similar fashion to what we did with the histograms.

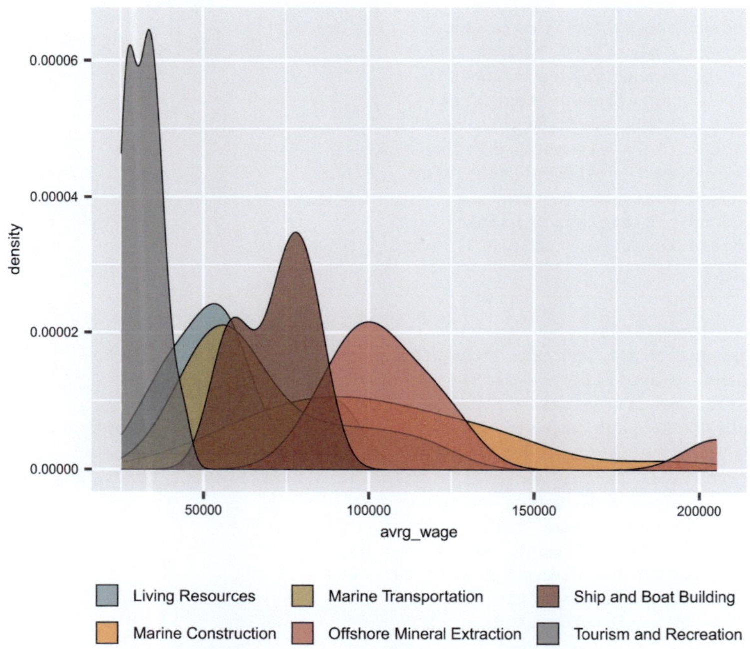

Fig. 2.6 Density curves of average wage by sector

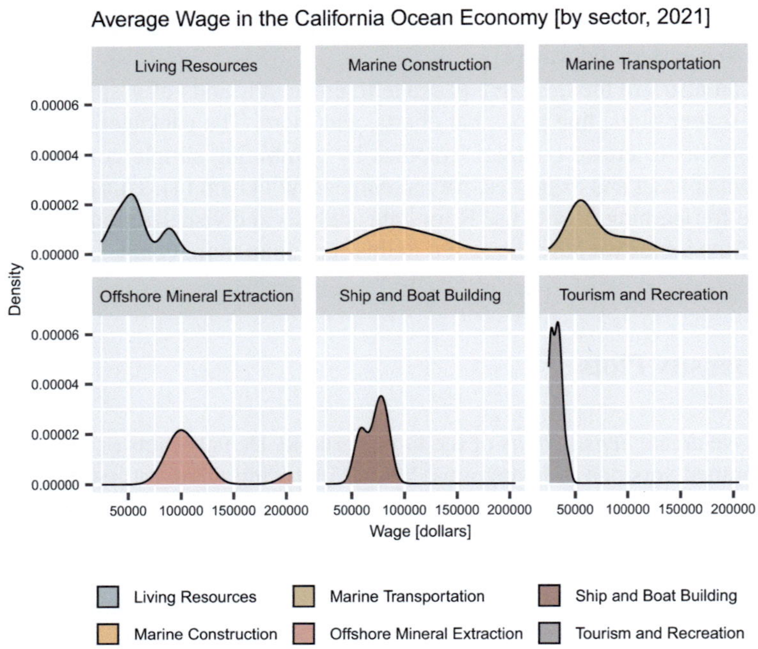

Fig. 2.7 Density curves of average wage by sector, using facets

```
options(scipen=10000)
ggplot(OceanEcon_CA, aes(avrg_wage, fill = sector)) +
  geom_density(alpha = 0.7, lwd=0.2) +
  scale_fill_manual(values = SoWest) +
  theme(axis.text=element_text(size=5),
        axis.title=element_text(size=6),
        plot.title = element_text(size = 8),
        legend.text = element_text(size = 6),
        legend.position="bottom",
        legend.title=element_blank(),
        legend.key.size = unit(0.3, "cm")
)
```

```
options(scipen=10000)
ggplot(OceanEcon_CA, aes(x = avrg_wage)) +
  geom_density(aes(fill = sector), alpha=0.5, lwd=0.3) +
  scale_fill_manual(values = SoWest) +
  labs(title="Average Wage in the California Ocean Economy [by sector, 2021]",
                  x="Wage [dollars]", y = "Density")+
  facet_wrap(~ as.factor(sector)) + theme(axis.text=element_text(size=5),
        strip.text = element_text(size = 6),
        axis.title=element_text(size=6),
        plot.title = element_text(size = 8),
        legend.text = element_text(size = 6),
        legend.position="bottom",
        legend.title=element_blank(),
        legend.key.size = unit(0.3, "cm")
)
```

2.7 Bonus Section: Scatterplots

Up until this point, we have created plots for a single variable. Similarly, we can use a different visualization technique when the objective is to represent the relationship between *two* metric variables. They are *scatterplots*, and we show one in Fig. 2.8. Although we have classified the data using different colors by sector, it is unclear how exactly these two variables (`average wage` and `average establishment size`) vary within each sector.

```
ggplot(OceanEcon_CA, aes(x=est_size, y=avrg_wage, color = sector)) +
     geom_point(size=2, shape=18) +
     scale_color_manual(values = SoWest) +
 labs(x="Average Establishment Size", y = "Average Wage") +
    theme(axis.text=element_text(size=6),
      strip.text = element_text(size = 6),
      axis.title=element_text(size=7),
      plot.title = element_text(size = 7),
      legend.text = element_text(size = 6),
      legend.position="bottom",
      legend.title=element_blank(),
      legend.key.size = unit(0.3, "cm")
)
```

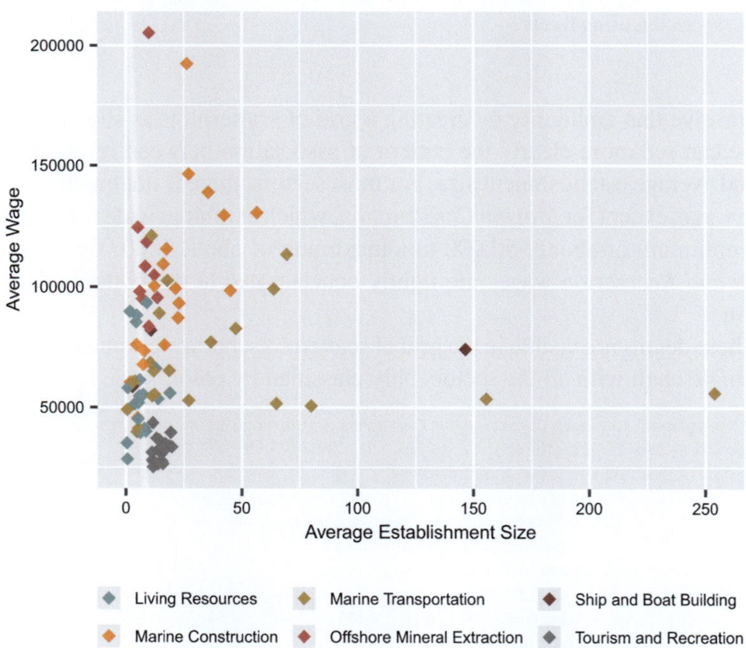

Fig. 2.8 Scatterplot of average wage and establishment size in the California Ocean Economy [by sector, 2021]

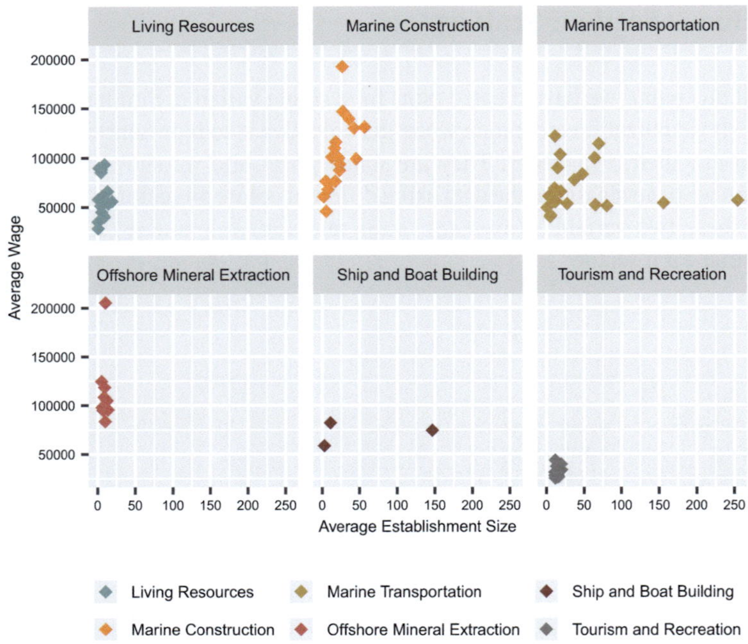

Fig. 2.9 Panel of scatterplots of average wage and establishment size in the California Ocean Economy, by sector, using facets

We resolve that ambiguity by creating a grid of scatterplots as shown in Fig. 2.9. Now we can see more clearly the pattern of association between levels of average wage and average establishment size. For most sectors, there is not much variation in average wage, except for *Marine Construction*, which exhibits a wide range of values, from a minimum of about \$50,000 to a maximum of about \$200,000. Conversely, *Tourism and Recreation* wages are tightly concentrated at the bottom of the wage spectrum.

All these details are visible in the grid of scatterplots, but were somewhat obscured in the single chart with all the sectors only classified by color.

```
ggplot(OceanEcon_CA, aes(x=est_size, y=avrg_wage, color = sector)) +
  geom_point(size=2, shape=18) +
  scale_color_manual(values = SoWest) +
  labs(x="Average Establishment Size", y = "Average Wage") +
  facet_wrap(~ as.factor(sector)) + theme(axis.text=element_text(size=5),
        strip.text = element_text(size = 6),
        axis.title=element_text(size=6),
        plot.title = element_text(size = 7),
        legend.text = element_text(size = 6),
        legend.position="bottom",
        legend.title=element_blank(),
        legend.key.size = unit(0.3, "cm")
)
```

References

1. Tufte, Edward R. 1983. *The Visual Display of Quantitative Information*. Cheshire, CT: Graphics Press. ISBN: 0-9613921-4-2.
2. Cleveland, William S. 1993. *Visualizing Data*. Summit, NJ: Hobart Press. ISBN: 0-9634884-0-6.
3. Few, Stephen. 2021. *Now You See It: Simple Visualization Techniques for Quantitative Analysis*, 2nd edn. Berkeley, CA: Analytics Press. ISBN: 1-9383771-2-5.
4. Wickham, Hadley. 2016. *ggplot2: Elegant Graphics for Data Analysis*. Springer, New York. ISBN: 978-3-319-24277-4. https://ggplot2.tidyverse.org.
5. Cleveland, William S., and Robert McGill. 1984. Graphical Perception: Theory, Experimentation, and Application to the Development of Graphical Methods. *Journal of the American Statistical Association* 79: 531–554.

Part II
Uncovering Relationships in the Data

Chapter 3
Hedonic Estimation Models

3.1 A Note on Regression Models in General

Regression modeling is one of the foundational tools in empirical research and applied analytics, particularly in fields where understanding relationships among variables is crucial. With origins dating back over 150 years—most notably formalized by Sir Francis Galton in his 1885 work—regression techniques have become indispensable in scientific, policy, and business domains. Their enduring popularity stems from their conceptual simplicity and computational tractability, allowing researchers and analysts to estimate the magnitude (and direction) of the effects of predictor variables (independent variables, IVs) on the outcomes of interest (dependent variables, DVs).

In the context of climate change adaptation, regression models are especially valuable. The impacts of climate variability are often diffuse, intersecting with socioeconomic, environmental, and infrastructural factors in complex ways. Regression models offer a structured way to isolate these relationships and provide evidence-based insights for decision-making. Whether estimating how precipitation levels influence flood risks or how air quality affects housing markets, regression techniques enable practitioners to quantify and interpret these linkages rigorously and transparently.

This chapter is not intended as a comprehensive treatment of regression analysis—a task well-covered by texts such as Gelman and Hill [1]. Instead, our goal is to introduce applied regression techniques through two targeted case studies, each designed to reflect practical scenarios encountered in climate change adaptation work. Our focus is on illustrating key ideas, demonstrating their implementation in R, and highlighting best practices in model building and interpretation.

The first case study uses a real, albeit somewhat dated, dataset to analyze the effect of environmental variables—namely, geographic location and air pollution levels—on housing prices. This example demonstrates the use of multiple linear regression, where the outcome variable (housing prices) is continuous and predictors

F. DePaolis, *Workflows for Climate Change Adaptation Analytics*,
https://doi.org/10.1007/978-3-032-14004-3_3

include a mix of categorical and continuous variables. This scenario is illustrative of common policy and planning questions where socio-environmental drivers need to be considered jointly.

In the second case study, we shift focus to a simulated dataset that assesses the likelihood of flooding as a function of precipitation levels. Here, the outcome is binary (flood/no flood), and the model employed is the logistic regression. This example is relevant to climate risk assessments, emergency response planning, and infrastructure resilience analysis, where practitioners must translate climate-related data into actionable outcomes relevant to stakeholders and decision-makers.

While the models differ in their technical formulations—linear versus logistic—the underlying logic remains consistent: using observed (or simulated) data to make informed inferences about real-world processes, by isolating the effects of policy-related variables on the outcome variable. In both cases, we emphasize model interpretation, diagnostics, and the communication of results, all of which are essential for translating statistical outputs into actionable knowledge.

Crucially, regression modeling is not merely a statistical exercise—it is a key element of evidence-based workflows for climate adaptation. By embedding these models within broader analytical pipelines, we can enhance transparency, reproducibility, and the policy relevance of our work. To that end, all examples in this chapter are implemented using widely supported R packages and reproducible coding practices that align with the ethos of open and collaborative science.

3.2 Hedonic Estimation

Accurately quantifying the value of environmental goods and assessing the socio-economic impact of climate change requires robust statistical tools. One such powerful approach is regression modeling, with a particularly valuable subfield being hedonic regression. This section introduces regression modeling through the lens of hedonic techniques, using a classic dataset to ground the discussion: the Boston Housing Market dataset.

Although the foundational concepts of hedonic pricing can be traced back to Court [2] and Griliches [3], it was Sherwin Rosen [4] who formalized and popularized the use of hedonic models in economics. His work established a systematic method for estimating the value of differentiated goods—goods that are not identical and whose individual attributes significantly affect their prices.

Hedonic regression, also known as *hedonic demand theory*, is a revealed preference method.[1] It estimates the value or demand for goods whose prices are not directly observable, especially when attributes are bundled—as is the case with housing, environmental quality, and local amenities. Environmental economists have long relied on hedonic models to estimate the implicit price of environmental goods, such as clean air, green space proximity, and noise reduction.

---

[1] Later in the book, we discuss the differences between *revealed* and *stated* preferences.

An early and influential application is the 1978 work of Harrison and Rubinfeld [5], who used data from the 1970 U.S. Census in Boston to estimate the demand for clean air. Their model incorporated air quality metrics such as concentrations of nitrogen oxides (NO_x) and particulate matter, as well as neighborhood and housing characteristics. Their findings highlighted how environmental quality is capitalized into housing prices—a fundamental concept for assessing adaptation needs and market responses to climate risks. Since then, hedonic modeling has become central to real estate valuation, and it has also found diverse applications, including the analytics of price indices such as the U.S. Consumer Price Index (CPI) (Fixler et al. [6] and Graf [7]); novel economic valuations such as *surfonomics* in Santa Cruz, California [8]; and the impact of wind farms on property values in Germany [9].

More recently, the growing urgency of the climate crisis, particularly sea-level rise and extreme weather events, has elevated the importance of hedonic modeling. It is increasingly used to assess how physical climate risks and climate adaptation interventions (e.g., seawalls, zoning changes, resilience infrastructure) affect land and property values. For practitioners in climate adaptation, this makes hedonic techniques indispensable [10].

We are using Regression Models. Why?

Regression modeling provides a statistical framework to quantify relationships between variables. In the context of hedonics, we use regression to estimate how individual housing attributes—including locational, environmental, and socio-economic factors—contribute to overall property value. This allows us to decompose the price of a good into the value of its components; detect spatial and temporal patterns in value; predict outcomes under different adaptation scenarios; and support evidence-based planning and policy design.

The interpretability and flexibility of regression models make them particularly useful for practitioners aiming to analyze large datasets, communicate findings to decision-makers, and test adaptation hypotheses with empirical data.

3.3 Boston Housing Market dataset

The Boston Housing dataset—first used in Harrison and Rubinfeld [5]—remains one of the most cited examples in hedonic pricing literature. It contains housing and neighborhood characteristics for 504 census tracts from the 1970 U.S. Census in Boston. Due to its historical importance and pedagogical value, it has been incorporated into numerous statistical packages, including the `MASS` package in R.

However, the dataset has not been without controversy. It includes a variable representing the percentage of African-American residents in each tract and a composite index reflecting socio-economic class based on male workers' education and occupation. These variables have prompted criticism over their reinforcement of systemic racism and classism, raising important concerns about the ethical use of data and the potential for statistical models to inadvertently encode harmful biases. We encourage readers to explore the numerous discussions available both in social media outlets and the academic literature, and to reflect critically on how data—especially historical and demographic—can propagate inequalities if not handled thoughtfully. In this chapter, we use a modified version of the dataset that excludes those controversial variables.

The retained variables are:

- `medv`: Median value of owner-occupied homes (in thousands of 1970 U.S. dollars)
- `chas`: Binary variable indicating whether the tract borders the Charles River
- `nox`: Nitrogen oxide concentration (parts per 10 million)
- `rm`: Average number of rooms per dwelling
- `ptratio`: Pupil-teacher ratio by town
- `crim`: Per capita crime rate by town

These variables are selected to illustrate key concepts in regression modeling and their application in climate-related housing value estimation. For example, we will explore how environmental quality (`nox`) and location features (`chas`) influence property values.

Table 3.1 shows descriptive statistics, including the number of cases, the mean, the median, the standard deviation, the minimum, and the maximum for the variables of interest. We have created a subset of the Boston dataset to include only the variables we plan to use later, and in the order we want them. We create a list (called `vars`) to do exactly that. Following best practices, the list includes the dependent variable `medv`first, ensuring that it will appear at the top of any output that different functions will produce. The `sumtable` function in the `vtable` package (sometimes used as `sf()`) creates the table with the descriptive statistics selected in the list.

```
# Select variables by name for clarity
vars <- c("medv", "chas", "nox", "rm", "ptratio", "crim")
```

```
vtable::sumtable(Boston[vars],
  col.breaks = 6, digits = 3, col.align = c("l","r","r","r","r","r","r"),
  numformat = NA, factor.numeric = TRUE,
  summ = list(
    c('notNA(x)','mean(x)','median(x)','sd(x)','min(x)','max(x)')),
  summ.names = list(c('n','Mean','Median','St Dev.','Min','Max')),
  out = "kable"
)
```

Table 3.1 Summary statistics

Variable	n	Mean	Median	St Dev.	Min	Max
medv	506	22.533	21.2	9.197	5	50
chas	506	0.069	0	0.254	0	1
nox	506	0.555	0.538	0.116	0.385	0.871
rm	506	6.285	6.208	0.703	3.561	8.78
ptratio	506	18.456	19.05	2.165	12.6	22
crim	506	3.614	0.257	8.602	0.006	88.976

3.4 Correlation Matrix and Plot

Pearson's product-moment correlation coefficient (or simply, correlation coefficient) is a measure of the covariation of two metric variables. Table 3.2 displays the correlation coefficient for each pair of variables in the reduced dataset.

We can see that the strongest correlation is between home values and the number of rooms. This is somewhat intuitive as it is easy to see that larger homes (measured in number of rooms) command higher prices. The second highest correlation value is for the pupil-teacher ratio; again, this is negative as a higher number of students per teacher could be perceived as a predictor of lower school performance, and hence a less desirable neighborhood. In this case, the coefficient is negative, as home values would decline for locations with a perceived lower quality of education. The third correlation value is for the NO_x levels; again, this is negative as higher levels of contaminants would reduce home prices. The fourth highest correlation coefficient is between home values and the crime rate. This coefficient is also negative, as home values would decline for locations with higher crime rates. Finally, the variable that captures the effect of proximity to the river has the lowest correlation coefficient. It is positive because proximity to the Charles River in Boston was perceived at the time as a positive amenity.

```
# Compute correlation matrix for selected variables
mcor <- round(cor(Boston[vars]), 3)

# Set upper triangle and diagonal to NA for cleaner display
mcor[upper.tri(mcor, diag = TRUE)] <- NA

# Convert to data frame for kable
lower <- as.data.frame(mcor)
lower <- lower[-1,-6]

options(knitr.kable.NA = '')
```

```
kable(lower, caption = "Correlation Coefficients", digits = 3,
     format = "latex", booktabs = TRUE, linesep = "") |>
  kable_styling(latex_options = c("striped", "HOLD_position", "scale_down"))
```

The other cells in Table 3.2 show the correlation coefficients for all pairs of independent variables. This is important information because one of the regression modeling assumptions is that independent variables must be independent from each other.

Table 3.2 Correlation coefficients

	medv	chas	nox	rm	ptratio
chas	0.175				
nox	–0.427	0.091			
rm	0.695	0.091	–0.302		
ptratio	–0.508	–0.122	0.189	–0.356	
crim	–0.388	–0.056	0.421	–0.219	0.29

```
corrplot(mcor,
        method = "ellipse",
        type = "lower",
        col = brewer.pal(n = 10, name = "RdYlBu"),
        diag = FALSE,
        tl.col = "black", tl.cex = 0.7, tl.srt = 30,
        cl.cex = 0.6,
        addCoef.col = "black",
        number.cex = 0.6)
```

The plot in Fig. 3.1 is a *correlogram* and shows the direction and intensity of the relationship between pairs of variables in the dataset. A correlogram is a graphic representation of the correlation between variables. We see that positive correlations are shown in intensifying shades of blue, with negative correlations in corresponding shades of orange.

The reader must also keep in mind that, regardless of the strength of the correlation between two variables, it does not measure or imply (unidirectional) causality. This topic is commonly discussed in introductory statistics textbooks and we do not further elaborate on it here.

From the workflow perspective, the basic function used to create the correlogram is `corrplot`, with several parameters to define the colors, fonts, and shapes in the chart. Feel free to explore all the options available in the package documentation.

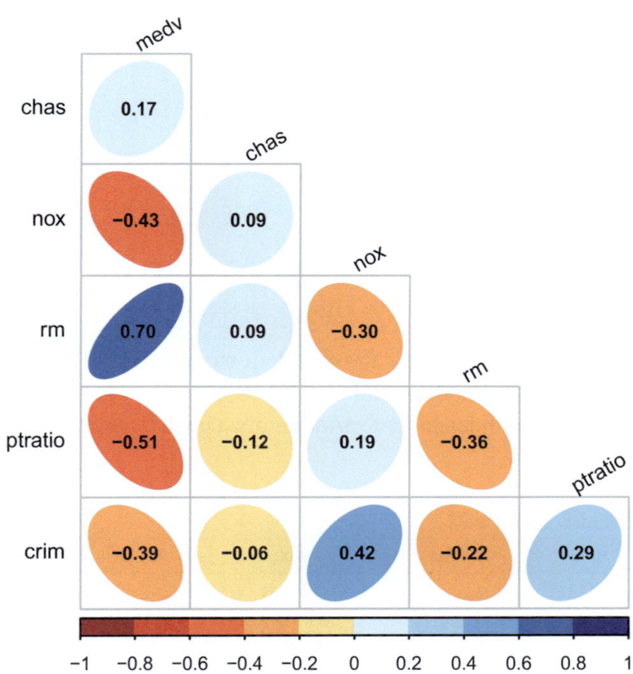

Fig. 3.1 Correlogram of selected variables

3.5 Scatterplots and Regression Lines

Using scatterplots (see the *Visualization* section) complements the numeric assessment of the relationship between the dependent variable and each of the independent variables we intend to include in the model. In Fig. 3.2, we show a grid of scatterplots with the dependent variable `medv` on the vertical axis, and each of the independent variables on the horizontal axis of each plot. Notice that the parameter `scales="free_x` allows the scale of the horizontal axis in each panel to adjust as a function of the values in the dataset for each variable. Also, notice that we shift from *wide* to *long* data formats.

```
# Formats the data for plotting shifting from wide to long format
Boston_long <- Boston[vars] |>
  gather(key, value, chas:crim)
```

```
# Scatterplots in facets
ggplot(Boston_long, aes(x = value, y = medv, color = key)) +
  geom_point(size=0.75) +
  geom_smooth(method = "lm", se = FALSE, lwd=0.5) +
  facet_wrap(~ key, scales="free_x", strip.position = "bottom") +
  theme(legend.position = "none",
        axis.text=element_text(size=5),
        strip.text = element_text(size = 6),
        axis.title.y = element_text(size=6),
        axis.title.x = element_blank()
        )
```

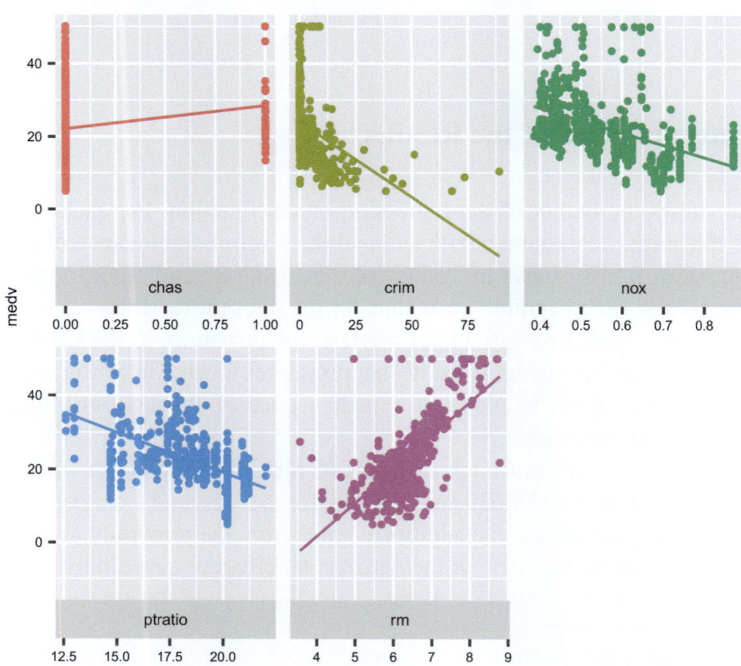

Fig. 3.2 Panel of scatterplots with best fit lines

The combination of numeric measures, such as correlation coefficients, and visual methods, such as correlograms and scatterplots with best-fit lines, serves as an additional component of the exploratory analysis that precedes the building of regression models. They are effective check-points that can help us anticipate potential problems such as a high correlation between independent variables (also known as *multicollinearity*), non-linear relationships that may require the consideration of alternative functional forms, and the presence of outliers, which may foreshadow poor model performance [11].

3.6 Building the Regression Model

The model below has the median home value as the dependent variable, and the proximity to the Charles River and NOx as the factors in the hedonic estimation. The other variables function as controls in the regression model.

```
hedonic.1 <- lm(medv~chas+nox+rm+ptratio+crim, data= Boston)
```

The `stargazer` function we use in the code block below is a useful way of extracting values from modeling result objects and presenting them in elegant tables.

```
stargazer(hedonic.1, ci=FALSE, type = "latex", header = FALSE,
        no.space = TRUE, single.row = TRUE, report = 'vc*',
        title= "Summary of the regression model results",
        label = "tab:reg-results")
```

3.7 Interpretation of the Results

Table 3.3 shows the output of the regression model function. For each variable, it displays the partial regression coefficient and the level of significance of the corresponding *t-test*. It also shows the values of $adjusted{-}R^2$ and other model performance measures. Let's examine the results for the relevant variables.

With respect to the first variable of interest, the proximity to the Charles River (`chas`), we see from the model's results that properties located in census tracts adjacent to the Charles River would have a median price of about $3,940 higher than properties located elsewhere.

For the second variable of interest (levels of NO_x), we see that it would reduce the values of the homes by $5,630 and $12,728, for the least and most polluted areas, respectively.[2] In other words, the additional value from the proximity to the Charles River is only about 30 and 70% of the detrimental effect of air pollution. Note that

[2] These values correspond to multiplying the minimum (0.385) and maximum (0.871) values of the variable nox and the corresponding partial regression coefficient (14.625).

Table 3.3 Summary of the regression model results

	Dependent variable:
	medv
chas	3.940***
nox	−14.625***
rm	6.775***
ptratio	−1.022***
crim	−0.130***
Constant	7.119*
Observations	506
R^2	0.627
Adjusted R^2	0.623
Residual Std. Error	5.645 (df = 500)
F Statistic	168.118*** (df = 5; 500)

Note *p<0.1; **p<0.05; ***p<0.01

each additional room would add $6,775 to the estimated price of the home. Also, keep in mind these are dollars from 1970!!

The value of $adjusted-R^2$ (0.623) indicates that this regression model explains about 62% of the variability in median home prices at the census tract level. The other 38% remains unaccounted for and could be explained by other variables in the dataset (but not currently included in the model), unobserved attributes (variables not in the dataset), or simply the random variability in prices.

3.8 Diagnostics

The theory behind regression modeling contains many assumptions and constraints. Fortunately, R provides the analyst with rather simple tools to verify whether those assumptions are met, or, if they are not, the magnitude of the departure from those assumptions. The simplest way is to get a diagnostic plot from the regression model itself. The `autoplot` function below produces a 4-panel chart, shown in Fig. 3.3.

```
par(mfrow = c(2, 2))
plot(hedonic.1, col = "royalblue") +
  theme(axis.title=element_text(size=7)) +
  theme(title=element_text(size=9))
```

```
par(mfrow=c(1,1), fg="black")
```

```
## NULL
```

Like many other analytical techniques, linear regression makes several assumptions about the data, namely:

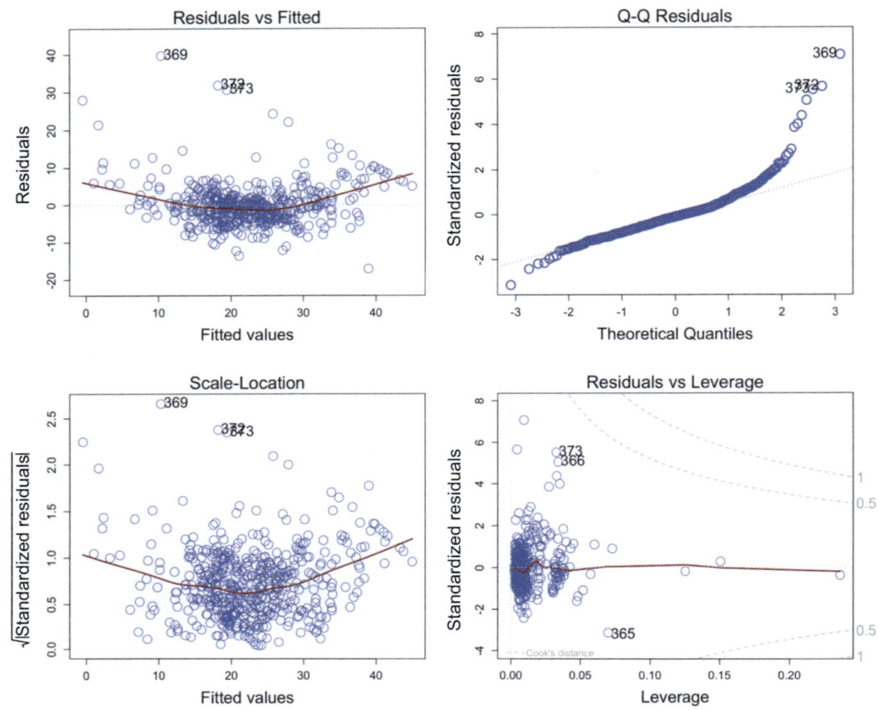

Fig. 3.3 Diagnostic plots (default configuration)

1. *Linearity*. The relationship between the predictor (x) and the outcome (y) is assumed to be linear.
2. *Normality of residuals*. The residual errors are assumed to be normally distributed.
3. *Homogeneity of the variance of residuals*. The residuals are assumed to have a constant variance (homoscedasticity)
4. *Independence of residual error terms*. That is, the distribution of residuals should not display any particular pattern as a function of the level of the predicted values.

You should check whether or not these assumptions hold true. Potential problems include:

1. *Non-linearity* of the outcome–predictor relationships
2. *Heteroscedasticity*: Non-constant variance of error terms.
3. Presence of *influential values* in the data that can be:

    a. *Outliers*: extreme values in the outcome variable.
    b. *High-leverage points*: extreme values in the predictor variables.

All these assumptions and potential problems can be checked by producing some diagnostic plots visualizing the residual errors. For additional details, see Fox and

Weisberg [12], James et al. [13], and the STHDA's webpage on *Linear Regression Assumptions and Diagnostics in R: Essentials.*

The code below selects three of the four available charts (the default option of the `plot()` function) and plots them in a row. This alternative gives the analyst additional choices, such as more detailed labels and, if needed, different colors. The three panels are, from left to right, *Residual v Fitted values*; *Q–Q Residuals*; and the *Scale–Location* chart. These correspond to the upper-left, upper-right, and bottom-left, respectively, the default, 4-panel chart in Fig. 3.3.

The left panel in Fig. 3.4 shows the distribution of residuals. Under the regression assumptions, we would expect most points to be more or less uniformly distributed across the horizontal axis, within a narrow range of values around zero on the vertical axis. The red line goes through the average residuals at each level of the fitted values. In our case, the line is located at a higher level for both low (below $10,000) and high (above $35,000) fitted values, and at or very close to zero for values in the middle of the scale. That means that the model *underestimated* low and high values compared with the observed data.

The other two panels in Fig. 3.4 reflect similar model performance issues, but now through the analysis of *standardized* residuals. The residuals have been converted into standard deviation units. In cases where the analysis requires building several models, considering measures based on standardized residuals facilitates the comparison among models that use variables measured in different units.

All three panels in Fig. 3.4 are very efficient in displaying the information analysts need to determine whether the model meets the requirements of regression assumptions.

```
par(mfrow=c(1,3), fg="white")

# Zero mean of residuals
plot(hedonic.1, which=1, col="royalblue")
box(col="dimgray"); axis(1, col="dimgray", col.ticks="dimgray"); axis(2, col="
    dimgray", col.ticks="dimgray")
title(xlab="Fitted Values", ylab="Residuals", main="Zero mean of residuals")
```

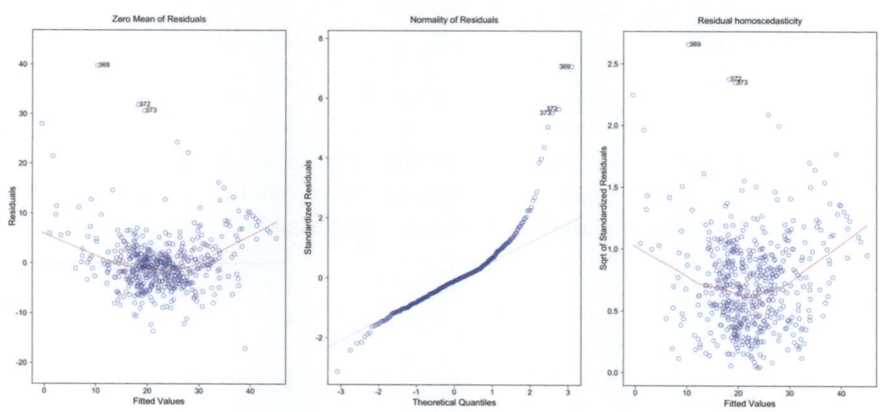

Fig. 3.4 Selected diagnostic plots

```
# Normality of residuals
plot(hedonic.1, which=2, col="royalblue", ann=FALSE, sub=" ")
box(col="dimgray"); axis(1, col="dimgray", col.ticks="dimgray"); axis(2, col="
    dimgray", col.ticks="dimgray")
title(xlab="Theoretical Quantiles",ylab="Standardized Residual",main="Normality
     of residuals")

# Homoscedasticity (Constant residual variance)
plot(hedonic.1, which=3, ann=FALSE, sub=" ", col="royalblue")
box(col="dimgray"); axis(1, col="dimgray", col.ticks="dimgray"); axis(2, col="
    dimgray", col.ticks="dimgray")
title(xlab="Fitted Values",ylab="SqRt of Standardized Residual", main="Residual
     homoscedasticity")
```

```
par(mfrow=c(1,1), fg="black")
```

3.9 Which Variable has the Strongest Impact on the Predicted Values?

The interpretation of the partial regression coefficients discussed above addresses the impact that each variable has on the predicted values of the dependent variable. For each variable, the coefficients represent the changes in the dependent variable, `medv`, for each unit change in the corresponding independent variable. Now the question is, can we use those magnitudes to determine which variable is more influential? The simpler answer is *no, not directly*. The variables in the model have different units and scales, and we cannot derive the relative effect of each of them from the partial regression coefficients themselves. We have to estimate the 'standardized regression coefficient' and use those to determine which variable has the strongest effect on the dependent variable. The standardized regression coefficient is also called 'beta-weights' (β-weights).[3] The R function `lm` doesn't include the β-weights in its output. There are, however, at least two ways of calculating them. The first approach consists of manually calculating them using the equation:

$$\beta_{x_i} = b_{x_i}\left(\frac{S_{x_i}}{S_y}\right) \tag{3.1}$$

where β_{x_i} is the standardized regression coefficient for the ith independent variable; b_{x_i} is the regression coefficient for the ith independent variable; and S_{x_i} and S_y are the standard deviations of the ith independent variable and the dependent variable, respectively.

[3] Some pieces of software, like SPSS, include them in the regression modeling results by default.

The second approach is based on estimating a regression model with the standardized versions of all the variables. The resulting regression coefficients will be the standardized regression coefficients.

The interpretation of the standardized regression coefficients is that they represent the change in the dependent variables, measured in standard deviation units, when the corresponding independent variable changes by one standard deviation. Because the variables have been transformed to their standardized version (*z-scores*, if you will), the effects of scales and units have been removed. That way, we can use the regression coefficients to determine which variables are more influential in the model. Higher values (either positive or negative) would have a stronger positive or negative influence on the predicted value of the dependent variable.

3.10 Standardized Regression Coefficients

```
# Define variables of interest
vars <- c("chas", "nox", "rm", "ptratio", "crim")

# Calculate SD ratios for beta-weight formula
sd_ratios <- sapply(vars, function(v) sd(Boston[[v]]) / sd(Boston$medv))

# Extract regression coefficients (excluding intercept)
coeffs <- coef(hedonic.1)[vars]

# Calculate standardized coefficients (beta-weights)
betas <- coeffs * sd_ratios
```

The code above computes the standardized coefficients using Eq. (3.1). The next chunk of code prints the table with both the partial regression coefficients and the corresponding $\beta-$*weights*. You can see in Table 3.4 that the highest partial regression coefficients do not necessarily correspond to the highest $\beta-$*weights*. This is a common source of confusion, and hopefully our explanation in this section clarifies the application of this transformation.

```
# Create summary table
betas_table <- cbind(b = coeffs, beta = betas)
rownames(betas_table) <- vars
```

Table 3.4 Regression coefficients and beta-weights

	b	beta
chas	3.940	0.109
nox	–14.625	–0.184
rm	6.775	0.518
ptratio	–1.022	–0.241
crim	–0.130	–0.121

```
kable(betas_table, caption = "Regression Coefficients and Beta-weights", digits
      = 3,
    col.names = c("b", "beta"),
    format = "latex", booktabs = TRUE, linesep = "") |>
  kable_styling(latex_options = c("striped", "HOLD_position", "scale_down"))
```

Table 3.4 shows both the partial regression coefficients and the $\beta-weights$. We can see that the most influential variable, according to the level of the $\beta-weights$, is the number of rooms, and the second most important is pupil-teacher ratio. Proximity to the Charles River ranks last in the level of influence on home prices, while NO_x is the third highest. As we can see, the partial regression coefficients do not capture this order for the reasons we discussed earlier.

However, to some extent, using standardized coefficients brings us back to correlation coefficients (Table 3.2). A careful observation of the coefficients indicates that the order of intensity of correlation coefficients (absolute values) does indeed show the same order as the standardized partial regression coefficients.

3.11 Policy Implications

In this section, we use the results of a hedonic regression model to extract the intrinsic value of clean air or the proximity to the Charles River. In recent years, researchers have explored the potential effects of sea level rise (*SLR*) on real estate markets. For the most part, it seems, markets have not broadly internalized the higher risks associated with SLR. In most US markets, values have not declined for properties located on shorelines [14]. If that were the case, when using models like the one in this section, the coefficient for the proximity amenity would have a declining value over time, or even a negative sign. This would be the manifestation of buyers' preferences for properties located away from the water edge as captured by lower prices for properties adjacent to a perceived risky feature.

Unfortunately, this has not yet materialized. The risks are very real and of significant size in terms of the number of people and housing units affected, as discussed in Bernstein et al. [15], Yu [16], Murfin et al. [17], Gourevitch et al. [18], and McNamara et al. [19]. A combination of market forces and delayed policies (especially those related to zoning and insurance [14] conspires to facilitate the necessary market adjustments to internalize the true magnitude of climate impact effects [20]. Note that Bernstein et al. [15] found that there is some decline in price due to the expected effects of SLR.

To a large extent, the ability of local governments and residents to prepare and adapt to the climate crisis is determined by the market signals available to them. If those signals are muffled by policy decisions made elsewhere or short-term speculative market practices, the efficacy of adaptation decisions will be severely curtailed. Analysts can make significant contributions to climate change adaptations by applying relevant methods to appropriate data to inform the decision-making process.

3.12 Closing Comments

In closing, we want to remind the reader that the dataset we used in this section is over 55 years old. Many changes have taken effect in Boston since 1970. For instance, although automotive traffic—and associated pollution—has certainly increased, pollution from other sources has declined dramatically. Industrial operations have moved away from the region or become significantly cleaner. Other common features of mid-century American cities, like incinerators and railroads, have all but disappeared, either because of technological change or regulatory developments. Gentrification has taken hold of important areas of the city, and preferences for urban amenities have shifted for younger generations.

All these changes most likely have altered the preferences for living closer to the river, compared with older views. It is quite likely that if we run the same hedonics model with a contemporary dataset, we would obtain a much higher effect of the proximity to the Charles River, and air pollution may be of much lower concern for home buyers.

References

1. Gelman, Andrew, and Jennifer Hill. 2006. *Data Analysis Using Regression and Multilevel/hierarchical Models*. Cambridge University Press.
2. Court, A.T. 1939. *Hedonic Price Indexes with Automotive Examples*. Technical Report, pp. 99–117.
3. Griliches, Zvi. 1961. Hedonic price indexes for automobiles: An econometric of quality change. In: *The Price Statistics of the Federal Government*. NBER Chapters. National Bureau of Economic Research, Inc., pp. 173–196.
4. Rosen, Sherwin. 1974. Hedonic prices and implicit markets: Product differentiation in pure competition. *Journal of Political Economy* 82(1): 34–55. https://doi.org/10.1086/260169.
5. Harrison, David Jr., and Daniel L. Rubinfeld. 1978. Hedonic housing prices and the demand for clean air. *Journal of Environmental Economics and Management* 5(10): 81–102.
6. Fixler, Dennis J. et al. 2000. The use of hedonic regressions to handle quality change: The experience in the U.S. CPI. https://api.semanticscholar.org/CorpusID:62820949.
7. Graf, Brian. 2020. *Consumer Price Index Manual, 2020: Concepts and Methods*. USA: International Monetary Fund. https://doi.org/10.5089/9781484354841.069. ISBN: 9781484354841.
8. Scorse, Jason, Frank Reynolds III, and Amanda Sackett. 2015. Impact of surf breaks on home prices in Santa Cruz, CA. *Tourism Economics* 21(2): 409–418. https://doi.org/10.5367/te.2013.0367. eprint: https://doi.org/10.5367/te.2013.0367.
9. Sunak, Yasin, and Reinhard Madlener. 2017. The impact of wind farms on property values: A locally weighted hedonic pricing model. *Papers in Regional Science* 96(2): 423–444. https://doi.org/10.1111/pirs.12197.
10. Bishop, Kelly C. et al. 2020. Best practices for using hedonic property value models to measure willingness to pay for environmental quality. English (US). *Review of Environmental Economics and Policy* 14(2): 260–281. https://doi.org/10.1093/reep/reaa001. ISSN: 1750-6816.
11. Agresti, Alan, and Barbara Finlay. 2014. *Statistical Methods for the Social Sciences*, 4th edn. Harlow, England: Pearson. ISBN: 1-292-03489-0.
12. Fox, John, and Sanford Weisberg. 2019. *An R Companion to Applied Regression*, 3rd edn. Thousand Oaks CA: Sage. https://www.john-fox.ca/Companion/.

13. James, Gareth, et al. 2013. *An Introduction to Statistical Learning: With Applications in R*. Springer. https://www.statlearning.com/.
14. Keenan, Jesse M, Thomas Hill, and Anurag Gumber. 2018. Climate gentrification: From theory to empiricism in Miami-Dade County, Florida. *Environmental Research Letters* 13(5): 054001. https://doi.org/10.1088/1748-9326/aabb32.
15. Bernstein, Asaf, Matthew T. Gustafson, and Ryan Lewis. 2019. Disaster on the horizon: The price effect of sea level rise. *Journal of Financial Economics* 134(2): 253–272. https://doi.org/10.1016/j.jfineco.2019.03.013. ISSN: 0304-405X.
16. Yu, Edward. Dec. 2020. *Sea Level Rise and Its Impact on California Housing Markets*. UCLA Anderson Forecast: Technical Report.
17. Murfin, Justin, Matthew Spiegel, and Jose Scheinkman. 2020. Is the risk of sea level rise capitalized in residential real estate? *The Review of Financial Studies* 33(3): 1217–1255. https://doi.org/10.1093/rfs/hhz134.
18. Gourevitch, Jesse D., et al. 2023. Unpriced climate risk and the potential consequences of overvaluation in US housing markets. *Nature Climate Change* 13(3): 250–257. https://doi.org/10.1038/s41558-023-01594-8. ISSN: 1758-6798.
19. McNamara, Dylan E., et al. 2024. Policy and market forces delay real estate price declines on the US coast. *Nature Communications* 15(1): 2209. https://doi.org/10.1038/s41467-024-46548-6. ISSN: 2041-1723.
20. Hauer, Mathew E., et al. 2021. Assessing population exposure to coastal flooding due to sea level rise. *Nature Communications* 12(1): 6900. https://doi.org/10.1038/s41467-021-27260-1. ISSN: 2041-1723.

Chapter 4
Logistic Regression

4.1 Introduction

Logistic Regression is a special case of regression modeling in which the dependent variable is categorical, typically binary, but it could also accommodate multi-level categorical variables. At first glance, this seems to contradict the classic assumptions of linear regression, which require continuous variables and linear relationships between predictors and the response. However, through transformations of the variable space, namely the logit transformation, logistic regression adapts standard regression techniques to categorical outcomes, offering robust tools for prediction and classification tasks (Hosmer et al. [1]).

In this section, we use logistic regression in the context of climate change adaptation analytics to estimate the likelihood of flooding in urban neighborhoods under various rainfall scenarios. To that effect, we combine a *dichotomous*, or binary, variable (flood/no flood) with a continuous variable (rain in mm).

4.2 The Logic of Logistic Regression

Suppose we wish to estimate the probability π that a specific event occurs, for example, a street flooding after heavy rainfall. A naive approach might be to fit a linear model, such as $\pi = \alpha + \beta X$.

However, this is problematic because probabilities must lie between 0 and 1, while linear models can predict values outside this range. Moreover, in binary outcomes (e.g., flooded = 1, not flooded = 0), the linear regression often provides a poor fit, yielding low R^2 and high residual errors (in the form of high Root Mean Sum of Squared Errors [$RMSE$]). Figure 4.1 shows the case in which a linear regression model will poorly fit points representing a probability distribution.

F. DePaolis, *Workflows for Climate Change Adaptation Analytics*,
https://doi.org/10.1007/978-3-032-14004-3_4

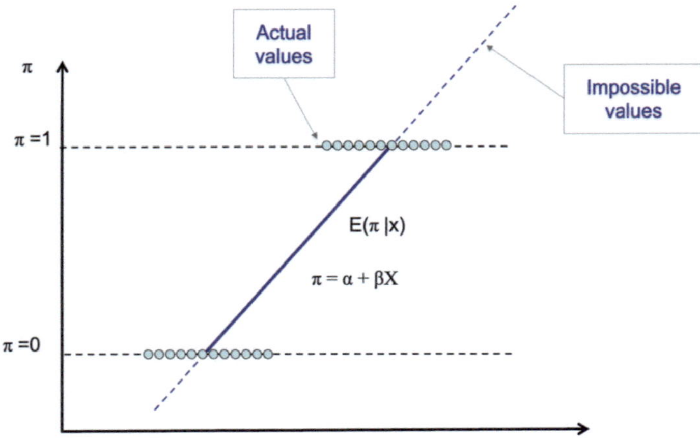

Fig. 4.1 Linear regression model

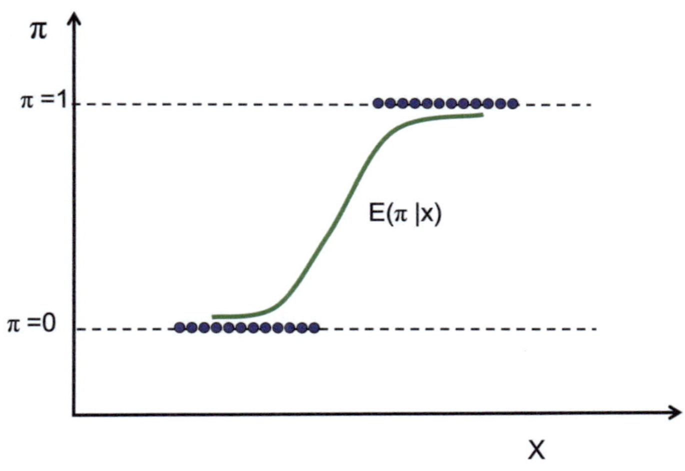

Fig. 4.2 Logistic regression model

To resolve this, the logistic regression approach uses a *sigmoid* (S-shaped) function to constrain the predicted probabilities to values between 0 and 1, like the one shown in Fig. 4.2. This is achieved via the logit transformation, $ln(\pi/1 - \pi) = \alpha + \beta X$, where the ratio $\pi/(1 - \pi)$ is called the odds. The logarithmic transformation of the odds is called *logit*, and its notation is $logit(\pi) = \alpha + \beta X$. This transformation models the log-odds of the event occurring as a linear function of predictors, where X can be a single vector representing one independent variable or a matrix representing multiple independent variables.

Interpretation of Model Components

The constant term, α, has a similar interpretation as the constant in a linear regression model. It is the value of the logit (natural log of the odds) when all the predictors (independent variables) are zero. We could use it to establish a baseline against which we measure the effects of the predictors.

The coefficient β in the logit model represents the change in the log-odds of the outcome per unit change in the predictor. For example, a positive β implies an increase in the probability of flooding as rainfall increases. The odds themselves are defined as:

Odds=$\pi/(1-\pi)$

So if $\pi = 0.75$, then $\pi/(1-\pi)$=0.75/0.25 and the odds of flooding are 3:1, meaning flooding is three times more likely than not-flooding.

Shape of the Logistic Function

- A large positive β steepens the S-curve near the midpoint (π=0.5).
- A negative β inverts the curve, modeling decreasing probabilities with increasing x.
- Smaller absolute values of β flatten the curve; that is, for similar changes in x, the probabilities change by a smaller amount.

Figure 4.3 shows how different values of β yield different logistic model shapes. Positive values of β produce forward S-shape lines, while negative values would produce backward shapes. In turn, the larger the absolute value of β, the steeper the central portion of the logistic line; and smaller absolute values of β would produce a flatter or stretched out curve.

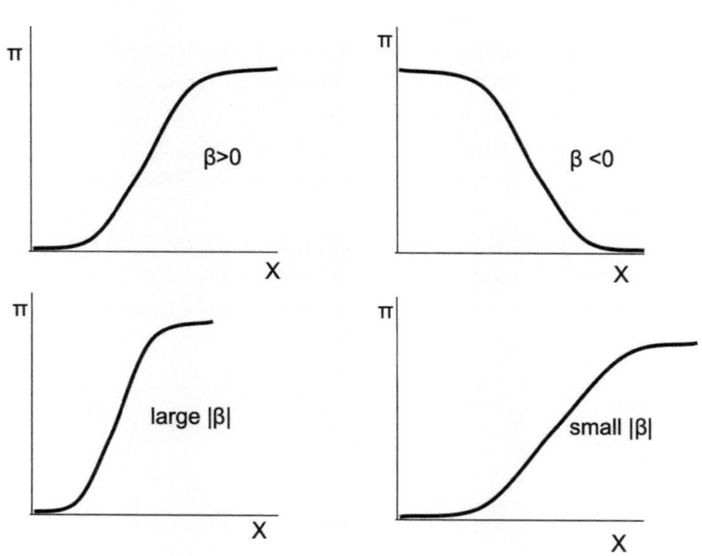

Fig. 4.3 Shapes of the logistic function

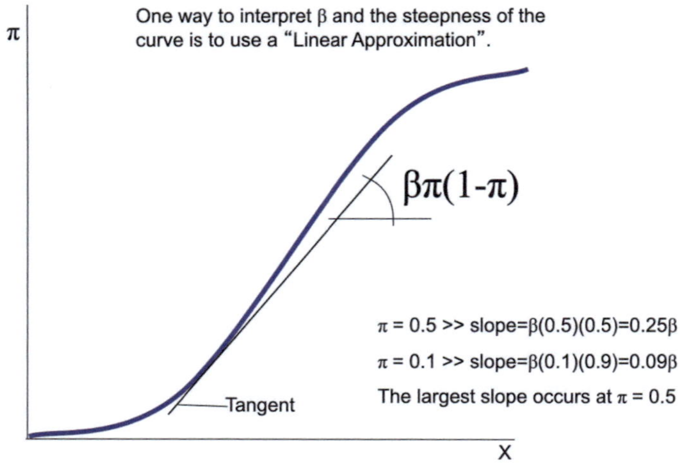

Fig. 4.4 Interpretation of the logistic regression coefficient

The tangent of the curve at any point π is given by $\beta\pi(1-\pi)$. From π=0.5, the tangent gets steeper and its slope increases in value, and then, after $\pi = 0.5$, the slope declines again, reaching zero again, at π=1. The maximum value of the slope of the tangent happens at π=0.5. Figure 4.4 shows all these conditions.

4.3 Data Setup

In this section, we use a simulated dataset that collects information on whether flooding[1] will occur in a particular neighborhood under different levels of precipitation (rain in mm). The data file is in native SPSS format (`flooding.sav`). We import it into R using the function `read.spss()` from the `foreign` package. Notice the parameter `to.data.frame=TRUE`. This ensures that the SPSS data file is properly read into an R data.frame.

Another way of importing a data file into R is by using the function `file.choose()` to locate the data file in the computer file system. This is a very efficient way of bringing up a data file into R without having to worry about coding the location of the file.

```
flooding <- foreign::read.spss("data/Flooding.sav", use.value.labels=TRUE,
                               max.value.labels=Inf, to.data.frame=TRUE)
```

[1] According to the Natural Resources Defense Council, urban flooding is intermittent inundation due to city landscapes that cannot absorb or otherwise manage rainfall.

4.4 Characteristics of Logistic Models

We use the `glm` function to fit a Generalized Linear Model with the `family= binomial(link="logit")` specification. This command will estimate the probability of an event happening as a function of the independent variables. Technically, the log-odds are a monotonic function of the probability or $ln(\pi/1-\pi)$ as we described above. Table 4.1 shows the results for this model.

```
### Build and displays the Logit Model
predict.flooding<-glm(Flood~Precipitation, family=binomial(link="logit"), data=
     flooding)
```

```
### Prints the output of the model
stargazer(predict.flooding, type = "latex", header = FALSE,
        ci=FALSE, no.space = TRUE, single.row = TRUE, report = 'vc*',
        title="Summary of Logistic Model Results",
        label = "tab:tabglm")
```

If we need the value of R^2, we can use the `lrm` function from the package `rms`. This is an extra step because the function `gml` does not generate the value on R^2. Table 4.2 shows the model results, including R^2=0.863. This indicates that about 86% of the variation in the probability of flooding is accounted for by precipitation. This indicates that the logistic model fits the observations very well.

Table 4.1 Summary of logistic model results

	Dependent variable
	Flood
Precipitation	0.217***
Constant	−12.323***
Observations	100
Log Likelihood	−17.090
Akaike Inf. Crit.	38.180

Note *p<0.1; **p<0.05; ***p<0.01

Table 4.2 Additional results using *lrm*

	Dependent variable
	Flood
Precipitation	0.217
Constant	−12.323
Observations	100
R^2	0.863
χ^2	103.808*** (df = 1)

Note *p<0.1; **p<0.05; ***p<0.01

```
### To get R-squared we use the rms packages
predict.flooding.2<-lrm(Flood ~ Precipitation , data=flooding)
#uses the "lrm" function from the rms package
# Nagelkerke's R-sq is under "Discrimination Indexes" as R2
```

```
stargazer(predict.flooding.2, type = "latex",header = FALSE,
        ci=FALSE, no.space = TRUE, single.row = TRUE, report = 'vc*',
        title="Additional results using $lrm$ ",
        label = "tab:tablrm")
```

4.5 Interpreting the Logistic Regression Model

We now have a logistic model that represents the relationship between the log of the odds of flooding (logit) and the amount of rain, such that

$ln\frac{\pi}{1-\pi} = -12.323 + 0.217x$; where x is the level of precipitation.

Let's interpret the components of this model, as well as some relevant points in the function it represents. First, the constant. We observe that α is relatively large in absolute value, 12.323; that means that the probability of flooding is extremely low when there is no rain (which seems quite intuitive). This is so because $e^{-12.323} = 4.448x10^{-6}$, and π is effectively zero.

Now the coefficient of the linear term, β, is the change in the log of the odds per unit change in precipitation. When precipitation goes up by 1mm, $ln\frac{\pi}{1-\pi}$ goes up by $e^{0.217}$, which is 1.2423. This indicates that the odds of flooding go up by 24.23% with each additional mm of rain.

Then we may want to know at what level of rain the chances of flooding go from *less* likely ($\pi < 0.5$) to more likely ($\pi > 0.5$). For that, we need to consider that the odds for $\pi = 0.5$ are 1, and the $logit(\pi) = 0$. It follows that $0 = \alpha + \beta x$; and $x = -\alpha/\beta$. Using the coefficients in the model, we obtain $12.323/0.217$ or 56.79mm. Above that level of rain, the chances of flooding are higher than 50%, and lower than 50% for any value below it.

Let's now consider two values in the vicinity of the 50% cutoff, say 50mm and 60mm, and compute the probability of flooding for each of them.

At 50mm

$odds = exp(-12.323 + 0.217(50)) = 0.2292$; this implies that π=0.1865 (and $1 - \pi$ =0.8135). There is only an 18.65% chance that 50mm of rain will produce flooding.

At 60mm

$odds = exp(-12.323 + 0.217(60)) = 2.0077$; this implies that π=0.6675 (and $1 - \pi$ = 0.3225). With a precipitation of 60mm, the chances of flooding are 66.75%.

When we present the interpretation in this fashion, it becomes more understandable and ready to be translated into actionable plans. The function is displayed in Fig. 4.6.

4.6 Diagnostics

As we have done in other sections, we need to assess the performance of the model. In this case, we use a combination of diagnostic tools. First, we get the variance-covariance matrix of the main parameters of the fitted model. This is a common feature of logistic regression analysis, and it contains information about each variable (the variance, on the diagonal of the matrix) and the relationship between pairs of variables (the covariance, on off-diagonal cells). The values are the unstandardized version of the correlation between variables. As we have seen in the linear regression model, independent variables should be strongly correlated with the dependent variable, and uncorrelated with each other.

Then we run the Wald test to confirm that the estimated coefficients are significantly different from the hypothesized value, in this , zero. The Wald test uses a χ^2 (chi-squared) distribution.

```
## Variance-Covariance matrix
vcov(predict.flooding)
### Wald Test
wt <- wald.test(b = coef(predict.flooding), Sigma = vcov(predict.flooding),
     Terms = 2)

wt
```

```
## (Intercept) Precipitation
## (Intercept) 8.2981625 -0.141613355
## Precipitation -0.1416134 0.002475588
## Wald test:
## ----------
##
## Chi-squared test:
## X2 = 19.0, df = 1, P(> X2) = 0.000013
```

The following step in the diagnostic process is building the *confusion matrix*, shown in Table 4.3. This is a table showing the actual values and the predicted values from the model. This gives us a measure of how many cases were correctly estimated. In the case of the flooding model, it correctly estimated 50 instances when flooding did not occur, and 44 instances when it did. In Table 4.4, the term *accuracy* captures this measurement, 0.94 or 94 out of 100. The other six cases are erroneous estimations.

```
### Extract the probabilities from the model
P=predict(predict.flooding, type="response")
### Rounds up/down the values of probabilities to 0/1.
p.flood=round(P)
### Creates the matrix of actual/predicted values;
```

Table 4.3 Confusion table

Prediction	Reference	
	0	1
0	50	2
1	4	44

Table 4.4 Associated Statistics

Term	Estimate	p.value
Accuracy	0.940	0.000
Kappa	0.880	
McNemar		0.683
Sensitivity	0.926	
Specificity	0.957	
pos_pred_value	0.962	
neg_pred_value	0.917	
Precision	0.962	
Recall	0.926	
f1	0.943	
Prevalence	0.540	
detection_rate	0.500	
detection_prevalence	0.520	
balanced_accuracy	0.941	

```
conf.matrix <- caret::confusionMatrix(as.factor(p.flood), as.factor(
     flooding$Flood))
```

```
kable(conf.matrix$table, caption = "Confusion Table", digits=3,
     format="latex", booktabs=T, linesep="") |>
  add_header_above(header = c("Prediction" = 1, "Reference" = 2),
                  align =c("l","r") ) |>
  kable_styling(latex_options = c("striped", "HOLD_position","scale_down"))
```

```
t.conf.matrix <- tidy(conf.matrix)

kable(t.conf.matrix[-c(2,4:5)], caption = "Associated Statistics", digits=3,
     format="latex", booktabs=T, linesep="") |>
  kable_styling(latex_options = c("striped", "HOLD_position","scale_down"),
               options(knitr.kable.NA = ''))
```

In addition, Table 4.4 (generated by the function `tidy(conf.matrix)`) shows a series of measures that can be useful depending on the specifics of the analysis. Not all these measures have direct application in analyses based on logistic regression, but many of them are more commonly used in statistical epidemiology and public health analytics.

```
          Reference
   Prediction 0 1
         0 A B A = True Positive B = False Positive
         1 C D C = False Negative D = True Negative
```

- **Metrics**
  - Accuracy = `(A+D)/(A+B+C+D)` Proportion of correct predictions

- kappa = Cohen's kappa statistic measures the agreement between the predicted values and the actual values.[2]
- McNemar = the p-value of the McNemar test.[3]
- Sensitivity = `A/(A+C)` Proportion of positive cases correctly predicted.
- Specificity = `D/(B+D)` Proportion of negative cases correctly predicted.
- Positive Predictive Value = `(sensitivity * Prevalence)/((sensitivity *Prevalence)+ ((1-specificity)*(1-Prevalence)))`
- Negative Predictive Value = `(specificity*(1-Prevalence))/(((1-sensitivity)*Prevalence)+((specificity)*(1-Prevalence)))`
- Precision = Positive Predictive Value.
- recall = sensitivity
- f1 = A score that combines the precision and recall measures.
- Prevalence = `(A+C)/(A+B+C+D)` Proportion of the true positive cases in the data.
- Detection Rate = `A/(A+B+C+D)` Proportion of true positive cases correctly predicted.
- Detection Prevalence = `(A+B)/(A+B+C+D)` Proportion of positive cases predicted by the model.
- Balanced Accuracy = `(Sensitivity+Specificity)/2`

4.7 Plots

Finally, we produce two chats. The first one (Fig. 4.5) shows the distribution of actual values for the level of precipitation and whether flooding occurred. The second one (Fig. 4.6) shows the predicted value for each level of precipitation.

```
### Plot actual data
colors_use <- c("seagreen", "indianred")
ggplot (flooding, aes(x=Precipitation, y=Flood)) +
  geom_point(aes(colour = factor(Flood)), size=1.5) +
  scale_colour_manual(name="Flooding", labels = c("No","Yes"), values =
        colors_use) +
  labs(title="Observed Flooding and Precipitation", x="Precipitation in mm", y=
        "Flooding [0=No, 1=Yes]") +
  theme(plot.title = element_text(size = 9),
       axis.title = element_text(size = 8),
       legend.title=element_text(size=7),
       legend.text=element_text(size=7))
```

```
### Plot the probability distribution
FloodProb<-data.frame(cbind(flooding$Precipitation, P))
ggplot (FloodProb, aes(x=V1, y=P)) +
  geom_point(aes(colour = P), size=1.5) +
  geom_line(aes(colour = P), size=0.75) +
  scale_colour_gradient(low = "seagreen", high = "indianred") +
```

---

[2] See details in the package documentation of the `confusionMatrix()` function.

[3] ibid.

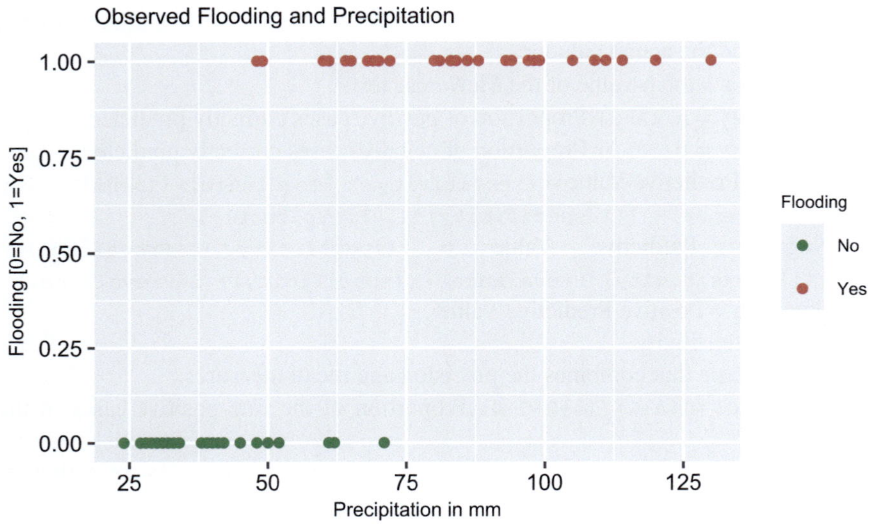

Fig. 4.5 Scatterplot of precipitation and flooding

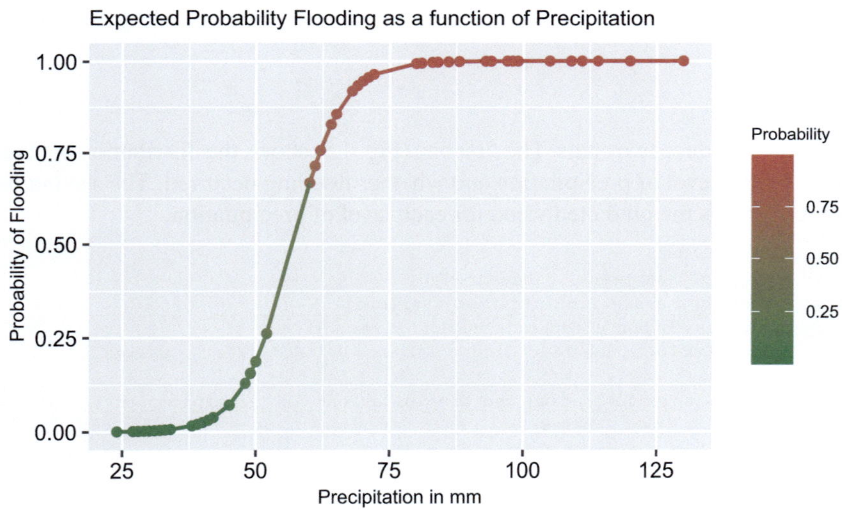

Fig. 4.6 Expected values from the logistic regression model

```
labs(title="Expected Probability Flooding as a function of Precipitation",
    x="Precipitation in mm", y="Probability of Flooding") +
      labs(colour = "Probability") +
theme(plot.title = element_text(size = 9),
     axis.title = element_text(size = 8),
     legend.title=element_text(size=7),
     legend.text=element_text(size=7))
```

4.8 An Application of Logistic Regression: Discrete Choice Models

Analysts frequently encounter the problem of assigning monetary value to environmental amenities. There are two main ways of collecting the information needed for such an estimation. One is to find out how much consumers effectively paid for environmental goods and services. For example, data from the National Park Service on the number of visitors, use and parking fees, fishing and hunting licenses, expenditures at licensed vendors, and other expenses would constitute the main source of data for what is called *revealed preferences*.

There are many amenities for which such data do not exist, either because it is not systematically collected, or because the markets for them do not exist. Where I live, California's Monterey Bay region, it is not unusual for people to park their vehicles on shoreline roads and contemplate the stunning natural beauty of the bay and its impressive wildlife, including sea otters, harbor seals, and, occasionally, whales. The public gets to enjoy these amenities without paying access fees, and sometimes, not even parking fees. There is, therefore, no easy way to impute a monetary value to what visitors experience.

In those cases, we use surveys to ask people how much they would be willing to pay to have access to particular places or for the enjoyment of certain experiences. These survey-based methods are used to elicit *stated preferences*, that is, those based on what people say, rather than what people do, or actually pay. It is possible to use data collected that way to assign a monetary value to transactions in a market that does not currently exist.

Within the stated preference methods, there are different specific approaches, such as *contingency valuation* (CV), *discrete choice experiments* (DCEs), *conjoint analysis*, and others. For the most part, all stated preference models are variations of logistic regression approach and its theoretical underpinnings. Logistic models are apt approaches in a variety of circumstances because they can model binary as well as multi-level response variables, and accommodate single or multiple independent variables. I point the reader to the work of Aizaki et al. [2] for excellent theoretical and operational discussion. The same team also developed and maintains the R package `DCchoice` [3].

4.9 A Note on Forecasting Under Uncertainty

In the Logistic Regression section, we utilized simulated data derived from a hypothetical flooding scenario. We want to highlight several emerging concerns that are becoming increasingly salient in environmental modeling. Whether one employs empirical observations from historical records or relies on simulated data, predictive estimates are invariably constrained by the context and assumptions under which the data were generated.

Consider the simulated flooding dataset: it lacks any temporal dimension associated with precipitation. We do not know whether the observations span the same year or several decades, or whether the data are recent or from a very distant past. Nor do we have information about the rate at which precipitation accumulated. For instance, did the 32 mm of rain—associated with no flooding—fall within 15 minutes or over the course of 9 hours? These unknowns render any causal inference speculative at best. Nonetheless, one might mistakenly conclude that 32 mm of precipitation is universally insufficient to trigger flooding at that location, an assumption that is both too simplistic and potentially misleading.

A more pressing concern, however, lies in the rapidly diminishing relevance of historical data in the face of accelerating climate change. While climate scientists may differ on the specifics, there is a broad consensus on a fundamental point: the climate is changing, and it will continue to change in non-uniform and unpredictable ways across regions. As Dr. Katharine Hayhoe aptly termed it, we are entering an era of "climate weirding" a state in which familiar patterns no longer hold, and novel, often extreme, phenomena emerge [4].

This introduces serious implications for predictive modeling. If future climatic regimes diverge substantially from those of the past, can we justifiably use historical data to estimate future events? An affirmative answer to this question would cast doubt on the validity of our models and the robustness of their forecasting capabilities. The response seems clear. It becomes essential to interrogate the assumptions under which historical data can still be utilized for modeling future scenarios. Foremost among these is a commitment to transparency: we must be explicit about our assumptions and methods, and ensure that our analyses are reproducible. This allows others to assess the impact of our design choices and, if desired, to formulate alternative approaches and compare outcomes.

4.10 Workflows in Regression Modeling

The workflow for regression modeling is very similar in all cases, regardless the particular type of application, and it is as follows:

1. Prepare the dataset depending on the type of dependent variable (metric or categorical)
   a. Determine if the models we plan to run need datasets in *wide* or *long* formats.
   b. Building a correlation table or chart is an effective complement to understanding relationships among variables.
   c. Transforming some variables into factors may be necessary to include them as indicator (*dummy*) variables.
2. Build the models (in R, there are different functions and parameters depending on the particular type of model we need to build)

3. Verify the significance of the model as a whole (using indicators such as R^2 and *F-test*)
4. Verify the significance of individual independent variables (*t-statistic*, *standard errors*, and *p-value*)
5. Determine the most influential variables (computing the $\beta-$weights)
6. Create diagnostic charts to determine if regression assumptions are met.

References

1. Hosmer, David W., Stanley Lemeshow, and Rodney X. Sturdivant. 2013. *Applied Logistic Regression*, 3rd edn. Wiley. https://doi.org/10.1002/9781118548387.
2. Aizaki, Hideo, Tomoaki Nakatani, and Kazuo Sato. 2014. *Stated Preference Methods Using R*. Boca Raton, FL: Chapman and Hall/CRC. https://doi.org/10.1201/b17292.
3. Nakatani, Tomoaki, Hideo Aizaki, and Kazuo Sato. 2023. *DCchoice: An R Package for Analyzing Dichotomous Choice Contingent Valuation Data*. R package version 0.2.0.
4. Hayhoe, Katharine. 2021. *Saving Us: A Climate Scientist's Case for Hope and Healing in a Divided World*. New York: Atria/One Signal Publishers.

Part III
Forecasting the Future

Chapter 5
Introduction to Time Series

Time series data structures consist of multiple observations recorded at regular time intervals. These datasets are prevalent in a variety of domains. For example, high-frequency financial data from NASDAQ—a stock exchange where prices and trading volumes are updated every second—enable analysts to monitor stock price fluctuations over the course of days, weeks, months, or even years. The vast volume of data generated supports the development of predictive models that can anticipate trends, both for individual stocks and for entire sectors.

Similarly, time series datasets have long been fundamental in climate science. Researchers have compiled data over the past 150 years, tracking variables such as atmospheric and oceanic temperatures, wind patterns, humidity, ocean salinity, and dominant wave periods. The oldest continuous climate record, from Central England, dates back to 1659.[1] This extensive historical record has facilitated the modeling of temperature changes associated with the concentration of greenhouse gases at various altitudes.

There are significant methodological differences between time series models and other statistical modeling techniques. Perhaps the most noticeable distinction is that time series models emphasize the temporal evolution of a single variable. Occasionally, analyses focus on the synchronicity of multiple variables. For further reading, see Cheong [1]. In contrast to cross-section (contemporaneous) regression models—which seek to identify causal relationships between independent and dependent variables—time series models prioritize the identification of patterns over time. The focus is on capturing the dynamics of a variable through its own historical behavior, rather than merely isolating its determinants.

The premise that future values can be projected solely from past observations is a powerful concept. Recognizing and modeling recurring temporal patterns offers substantial utility for data scientists, particularly in domains such as climate policy analysis. The ability to forecast without exhaustively accounting for every possible causal factor enhances the appeal and practical applicability of time series modeling.

[1] See https://www.metoffice.gov.uk/hadobs/hadcet/.

F. DePaolis, *Workflows for Climate Change Adaptation Analytics*,
https://doi.org/10.1007/978-3-032-14004-3_5

For the purposes of this chapter, we outline three principal modeling approaches:

1. Extrapolation Techniques: If the time series exhibits a relatively simple and well-defined trend, projection can be accomplished by fitting known functional forms (e.g., linear, exponential, or power functions) using regression techniques. This approach assumes that the underlying pattern will persist into the future.
2. ARIMA Models: For more complex time series—those exhibiting long-term trends, cyclical or seasonal behavior, and stochastic variability—we utilize specialized techniques. Cyclical and seasonal patterns are often conflated, but they differ: cycles do not have fixed frequencies, whereas seasonality is associated with calendar-based periodicity [2]. The AutoRegressive Integrated Moving Average (ARIMA) framework is a widely used method for handling such data, enabling decomposition and forecasting of intricate temporal dynamics.
3. Bayesian Time Series Models: Bayesian approaches are particularly useful when the fluctuations in the series do not have a clear cyclical behavior and stochastic effects dominate. After introducing a brief discussion on the distinctions between frequentist and Bayesian inferential frameworks, we demonstrate the application of Bayesian techniques to time series forecasting, providing enhanced capacity for uncertainty quantification and model updating.

5.1 Extrapolation Models

The goal in this section is to identify the 'functional form' of the dataset. That is a fancy way of saying that we want to find the equation that best describes all the values in the dataset. Then we want to use that equation to forecast future values, that is, those beyond the dataset we currently have. There are many possible equations at our disposal—maybe too many—but we will use linear (or straight line), and two curvilinear ones, the power function, which usually grows slowly, and the exponential, which typically grows much faster.

We are using Regression Models. Why?

We alluded to the possibility of using forecasting techniques in cases of time series that exhibit a definitive trend, and whose values display relatively small fluctuations. For this particular application, regression models are very efficient and relatively easy to set up for known functional forms that may match the data closely.

What follows is an adaptation of the regression techniques we learned in previous chapters. These are the main stages of the extrapolation process workflow:

1. Verify that the dataset follows the general shape of the assumption above
2. Visually determine the most likely shape of the dataset (for example, determining if the series is increasing at a constant rate or increasing at an increasing rate).
3. Match the shape to one or more known functional forms.
4. Obtain the function's parameters to build descriptive and predictive models.
5. Use that function to forecast the time series beyond the original dataset.

5.2 Sea Level Rise Dataset

For the examples in this section, we use a subset of an extensive dataset on global sea level rise available in Frederikse et al. [3]. The variable of interest is GMSL, global-mean sea level. Although the original dataset starts in 1900, we use a subset that starts in 1960. We also transformed the values by shifting the entire series upward so there are no negative values. We do this because some of the transformations in the models we will build require us to compute natural logarithms, and those are not defined for negative values. The dataset consists of two columns, one for the year and one for the value of GMSL. We labeled them `year` and `level`. Then we compute the logarithmic versions of those variables and named them `ln.year` and `ln.level`.

The code below reads the data and creates the new variables.

```
#Read dataset
Sea_level <- as.data.frame(read_csv("./data/GMSL-scale-1960.csv"))

# Variables needed for different models
Sea_level$ln.level <- log(Sea_level$level)
Sea_level$ln.year <- log(Sea_level$year)
```

As we have recommended earlier, it is a good practice to plot the data you plan to use. This is the most effective way of getting a clear understanding of the shape of the data, including overall distribution, any relevant distortions or clusters, as well as any extreme values. Figure 5.1 shows the observed values of the GMSL.

```
plot(Sea_level$year, Sea_level$level,
    type = "l", col = "royalblue", lwd = 1.5,
    xlab ="Year",
    ylab = "GMSL",
    cex.lab = 0.75,
    cex.axis = 0.5)
```

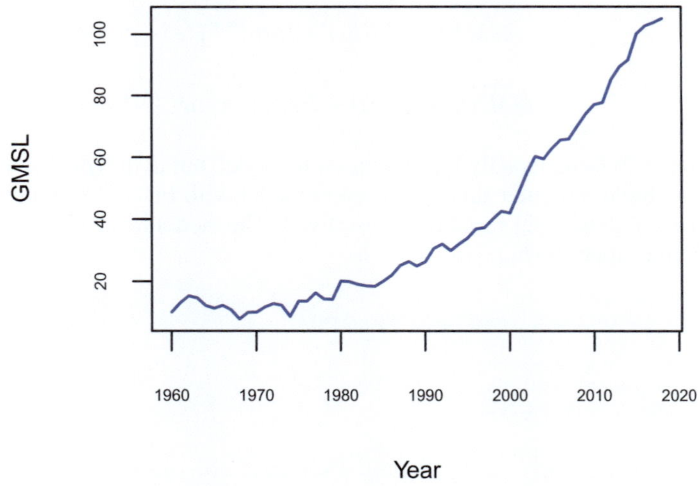

Fig. 5.1 Observed values of GMSL

5.3 Linear Fitting

For our fist model, we will use the variables in their original format (`year` and `level`) and the function `lm` to build the base model. The estimation equation is $G\hat{M}SL = b_0 + b_1 year$, where $G\hat{M}SL$ is the estimated value of $GMSL$; b_0 is the intercept; and b_1 is the slope. As we saw before, the algorithm in the function `lm` computes the regression coefficients such that the aggregate errors are minimized. More on that later. We create a new variable `level_hat.lin` to capture the values estimated by the model. We extract the regression coefficient by specifying the location of the values in the structure of the object created by the `lm` function (coefficients b_0 and b_1).

```
lin.model <- lm(level~year, data=Sea_level)
b_0 <- lin.model$coeff[1] # Intercept
b_1 <- lin.model$coeff[2] # Slope
Sea_level$level_hat.lin <- b_0 + b_1 * Sea_level$year # Estimated values y-hat
```

5.4 Exponential Fitting

Exponential models of the form $\widehat{GMSL} = b_0\, e^{b_1 year}$ are those in which the independent variable is an exponent. In order to be able to use a regular linear regression model, we must first convert the dependent variable $GMSL$ into its logarithmic format, $ln(GMSL)$. This is equivalent to building a linear regression model in a *log-linear* space. That is one in which the independent variable *year* remains in linear form, while the dependent variable, $GMSL$, is now in log format. Note that we use natural logs (base e).

To compute the logarithm, we combine the product and power rules for logs:

$$\widehat{GMSL} = ln(b_0) + ln(b_1^{year}) \tag{5.1}$$

$$\widehat{GMSL} = ln(b_0) + ln(b_1) \cdot year \tag{5.2}$$

Equation (5.2) looks exactly like a regression model, but in the *log-linear* space. Now we can build a regular linear regression model with $ln(GMSL)$ and *year* as dependent and independent variables, respectively. The model will yield the intercept, $ln(b_0)$, and the slope, $ln(b_1)$.

```
### Exponential model
exp.model <- lm(ln.level~year, data=Sea_level)

# Exponential model parameters
b_0.exp <- exp.model$coeff[1]
b_1.exp <- exp.model$coeff[2]

# Exponential model fitting
Sea_level$level_hat.exp <- exp(b_0.exp) * exp(Sea_level$year * b_1.exp)
```

In order to obtain the estimated value of the dependent variable, we use the function *exp* to revert to the original format.

5.5 Power Fitting

In this case, the independent variable is raised to a power, $\widehat{GMSL} = b_0 \ year^{b_1}$. Just like in the case of exponential models, we transform the variables to achieve the objective of estimating a regular, linear regression model in the appropriate space. In this case, a *log-log* space. Again, we apply the product and power rules of logs. $ln(GMSL) = ln(b_0) + b_1 \ ln(year)$; this equation now resembles a linear regression equation in a transformed space, in this case into a *log-log* space.

```
### Power model
pow.model <- lm(ln.level~ln.year, data=Sea_level)

# Power model parameter
b_0.pow <- pow.model$coeff[1]
b_1.pow <- pow.model$coeff[2]

# Power model fitting
Sea_level$level_hat.pow <- exp(b_0.pow) * (Sea_level$year^b_1.pow)
```

5.6 Assessing the Performance of Each Model

We now plot the observed values, as well as the values estimated by the linear, exponential, and power models, in a single chart. This is an efficient way of showing each model's fit to the observed values.

```
p1 <- p_alternative <- ggplot(Sea_level, aes(x = year)) +
  geom_line(aes(y = level, color = "Observed"), linewidth = 0.75) +
  geom_line(aes(y = level_hat.lin, color = "Linear"), linewidth = 0.5) +
  geom_line(aes(y = level_hat.pow, color = "Power"), linewidth = 0.5) +
  geom_line(aes(y = level_hat.exp, color = "Exponential"),
       linetype = 2, linewidth = 1.15) +
  labs(x = "Year", y = "Observed & estimated GMSL values") +
  scale_color_manual(
    name = "Values & Fittings",
    breaks = c("Observed", "Linear", "Power", "Exponential"),
    values = c("Observed" = "royalblue", "Linear" = "darkorange",
      "Power" = "magenta", "Exponential" = "green4" )
  ) +
  theme_minimal() +
  theme(
    panel.background = element_rect(fill = 'gray90', colour = 'gray46'),
    legend.text = element_text(size = 7),
    legend.title = element_text(size = 7),
    axis.title = element_text(size = 9),
    axis.text = element_text(size = 7)
  )

p1
```

The chart in Fig. 5.2 gives us a very good sense of which model is a better fit to the base dataset. Clearly, the linear model does not fit the observations very well.

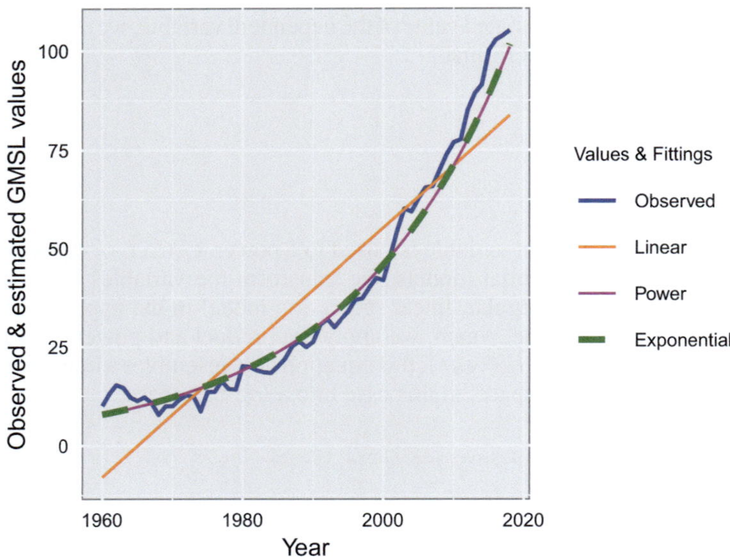

Fig. 5.2 Linear, exponential and power fitting

But then, there is something else. Both the *power* and *exponential* models fit the observations quite well…but we cannot distinguish between the two because the fitting lines are almost identical. This is not entirely unexpected. It is possible for a given dataset to have a combination of parameters that will yield a set of similar estimates for exponential and power functions.

As useful as they are, graphic methods are not precise enough to give us the certainty we need in assessing the goodness of fit of our models. For that, we need numeric measurements.

We use RMSE or root mean sum (of squared) errors to measure the aggregate differences between the estimates and the actual values. The equation is $RMSE = \sqrt{MSE}$; where $MSE = SSE/n$, and $SSE = \sum_{i=1}^{n} (y_i - \hat{y}_i)^2$ with y_i being the observed values; $\hat{y}_i$ the values estimated by each model; and n is the number of observations.

Table 5.1 shows the values of *RMSE* extracted directly from model results. The expression `lin.model$residual` extracts the vector of residuals from the linear model directly. The code then uses a nested structure to compute the rest of algorithm; it finally combines all the $RMSE$ values and displays the table with the results.

You can also compute them with the equations above, but the process is laborious, and the results would not change. We can see that the exponential model has an overall better *goodness of fit* because the measure of aggregate errors is the smallest of all

Table 5.1 Root mean sum of squared errors [RMSE]

Linear model	Exponential model	Power model
11.139	0.181	0.183

Table 5.2 R-squared for each of the extrapolation models

Linear model	Exponential model	Power model
0.8538	0.9453	0.9442

three models (although the power model is very similar). The RMSE values express numerically what we observed in the chart. The exponential and power models both perform very well, and the linear model produces large errors as it misses most of the observations by a significant amount.

```
# Extract RMSE values from model recults
rmse_results <- cbind(sqrt(mean(lin.model$residuals^2)),
                      sqrt(mean(exp.model$residuals^2)),
                      sqrt(mean(pow.model$residuals^2)))

colnames(rmse_results) <- c("Linear model","Exponential Model","Power Model")
```

```
# RMSE table
kable(rmse_results, caption = "Root Mean Sum of Squared Errors [RMSE]", digits
      = 3,
      format="latex", booktabs=T, linesep="") |>
  kable_styling(latex_options = c("striped", "HOLD_position","scale_down"))
```

The coefficient of determination, R^2, is a widely used measure of goodness of fit for regression models, and Table 5.2 shows the values for each model. In this case, we can also see that the lower the value of RMSE, the higher the value of R^2. Mathematically, this is because $R^2 = 1 - [df \cdot RMSE^2/SS_{Total}]$. We said that the power and exponential models are very similar; we have to show four decimal positions of R^2 to appreciate the difference between them.

```
# Extract R-squared from model results
R2.results <- cbind(summary(lin.model)$r.square,
                    summary(exp.model)$r.square,
                    summary(pow.model)$r.square)


colnames(R2.results) <- c("Linear model","Exponential Model","Power Model")
```

```
# R-squared table
kable(R2.results, caption = "R-squared for each of the extrapolation models",
      digits=4,
      format="latex", booktabs=T, linesep="") |>
  kable_styling(latex_options = c("striped", "HOLD_position","scale_down"))
```

5.7 Forecasting

Now that we have determined that the exponential model fits the data better than the others, we can use it to forecast future values, beyond the observations in the dataset. To that effect, we must first extract the coefficients $ln(b_0)$ and $ln(b_1)$ from the model results to compute the forecast values. We add the forecast values at the end of the observed values and produce a chart without discontinuities. First, we need to create two data objects, `year.future` and `level.future` to capture the values of the years beyond the dataset, and the estimated values of GMSL, respectively.

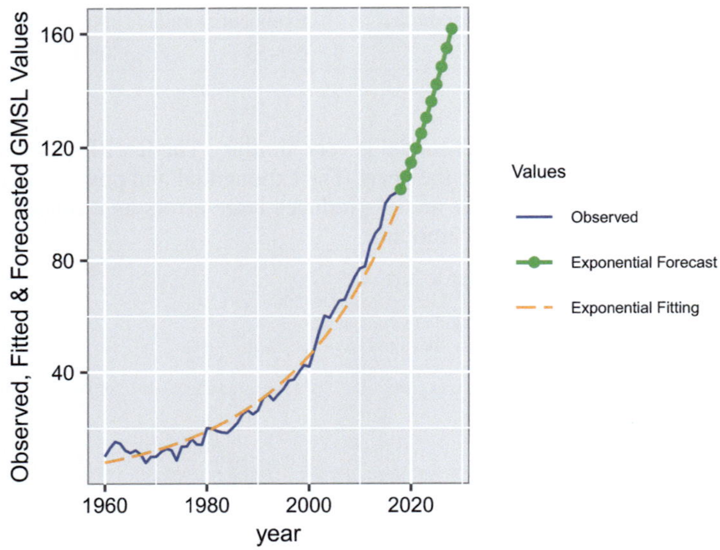

Fig. 5.3 Forecasting using exponential fitting

Table 5.3 Forecasted values using exponential fitting [GMSL in millimeters]

Year	Forecast
2018	105.04
2019	109.63
2020	114.43
2021	119.44
2022	124.68
2023	130.15
2024	135.88
2025	141.86
2026	148.11
2027	154.65
2028	161.48

```
# Forecasting
year.future <- seq(from = max(Sea_level$year),
                   to = max(Sea_level$year)+10, by = 1) # forecast 10 years into
                         the future
level.future.exp <- (tail(Sea_level$level, n=1)-tail(Sea_level$level_hat.exp, n
     =1)) + exp(b_0.exp) * exp(year.future * b_1.exp) # adds the first forecast
      value next to the last value in the original dataset
```

In Fig. 5.3, we plot the current data, the exponential model, and the extrapolated values corresponding to future time periods, in this case, 10 years into the future. We can also create a table with the estimated future values as shown in Table 5.3.

```
p2 <- ggplot() +
  geom_line(data = Sea_level, aes(x = year, y = level, color = "Observed"),
       linewidth = 0.5) +
  geom_line(data = NULL, aes(x = year.future, y = level.future.exp, color = "
       Exponential Forecast"), linewidth = 0.75) +
  geom_point(data = NULL, aes(x = year.future, y = level.future.exp, color = "
       Exponential Forecast"), linewidth = 1) +
  geom_line(data = Sea_level, aes(x = year, y = level_hat.exp, color = "
       Exponential Fitting"), lty = 5, linewidth = 0.5) +
  labs( x = "year", y = "Observed, Fitted & Forecasted GMSL Values") +
  scale_color_manual(
    name = "Values", breaks = c("Observed", "Exponential Forecast",
                          "Exponential Fitting"),
    values = c("Observed" = "royalblue", "Exponential Forecast" = "green3",
      "Exponential Fitting" = "orange" ) ) +
  theme(
    panel.background = element_rect(fill = 'gray90', colour = 'gray46'),
    legend.text = element_text(size = 6),
    legend.title = element_text(size = 7),
    axis.title = element_text(size = 9),
    axis.text = element_text(size = 8),
    legend.key = element_rect(colour = NA, fill = NA),
    legend.box.background = element_blank()
  )
p2
```

```
forecast <- cbind(year.future, level.future.exp)
colnames(forecast) <- c("Year", "Forecast")

kable(forecast, caption = "Forecasted values using exponential fitting [GMSL in
      millimeters]",digits=2,
    format="latex", booktabs=T, linesep="") |>
  kable_styling(latex_options = c("striped", "HOLD_position","condensed",
                              font_size=8))
```

A model with a very high coefficient of determination (R^2), as the one in the exponential model in this example, would, in general, give us a very high level of certainty. However, we must qualify that certainty to its real meaning for climate adaptation purposes. The high value of R^2 represents a good fit between the exponential function and the values in the dataset. However, those values have occurred at very different points in time, in this case between 1960 and 2020. We must ask ourselves how sure we are that future sea levels will change at the same rate as they did in that period. Most likely, we are not very confident that the values we forecast 10 years into the future will match the actual values. This is certainly something to take into account in our professional practice.

5.8 Closing Comments

We have followed the workflow introduced at the beginning of the section:

1. Verify that the dataset follows the general shape of the assumption above
2. Visually determine the most likely shape of the dataset (e.g., increasing at a constant rate or increasing at an increasing rate)
3. Match the shape to known functional forms (in this section, we are comparing linear, exponential, and power functions)

4. Obtain the function's parameters to build descriptive and predictive models.
5. Use that function to forecast the time series beyond the original dataset

We built a regression model using an exponential function as the reference (log-log formulation) and obtained the parameters of a model that fits the data very closely. We then used those parameters to extrapolate the dataset 10 years into the future. Even with the caveats introduced at the end of the section, we can say that we succeeded in producing a plausible estimate of future values.

References

1. Cheong, Jin. 2019. *Four ways to quantify synchrony between time series data*. https://medium.com/data-science/four-ways-to-quantify-synchrony-between-time-series-data-b99136c4a9c9.
2. Hyndman, Rob J., and George Athanasopoulos. 2018. *Forecasting: Principles and practice*. English. 2nd. Australia: OTexts.
3. Frederikse, Thomas, et al. 2020. The causes of sea-level rise since 1900. *Nature* 584(7821): 393–397. https://doi.org/10.1038/s41586-020-2591-3.

Chapter 6
ARIMA Models

6.1 Introduction

In this section, we discuss AutoRegressive Integrated Moving Average (ARIMA) models, a class of statistical models commonly used in time series forecasting. ARIMA models are particularly well-suited for data that exhibit non-stationarity, where trends and patterns evolve over time.

In the previous section, we employed regression models to fit time series data with the aim of forecasting—or, more precisely, extrapolating—values beyond the available dataset. However, this approach has limitations. Specifically, regression models assume a known functional form and are often inadequate when the data exhibit complex, non-linear trends or multiple seasonal components. Under such conditions, forecasts generated using regression models may yield high prediction errors.

To address these limitations, we now explore alternative modeling strategies. First, we will use built-in functions from the `forecast` package in R to produce a first-approximation forecast. Next, we will fit more specialized time series models, evaluating their suitability through both manual and automated model selection processes. Finally, we will employ the Holt-Winters exponential smoothing method, which is particularly effective for series with trend and seasonal components, to compare its performance with ARIMA-based models [1].

6.2 The Housing Permits Dataset

Urban municipalities face significant challenges as they adapt land-use practices to a changing climate. Among the most consequential of these practices is *managed retreat*, whereby existing development rights are revoked or modified to reflect evolving climate risks. This process may involve rezoning previously residential areas as open space, with the intent of discouraging permanent structures in hazard-prone zones. The implications of such policies are far-reaching: single-family residential zones, in particular, are difficult to adapt due to their embeddedness in social, economic, and cultural systems. Even modest regulatory changes in these areas may

F. DePaolis, *Workflows for Climate Change Adaptation Analytics*,
https://doi.org/10.1007/978-3-032-14004-3_6

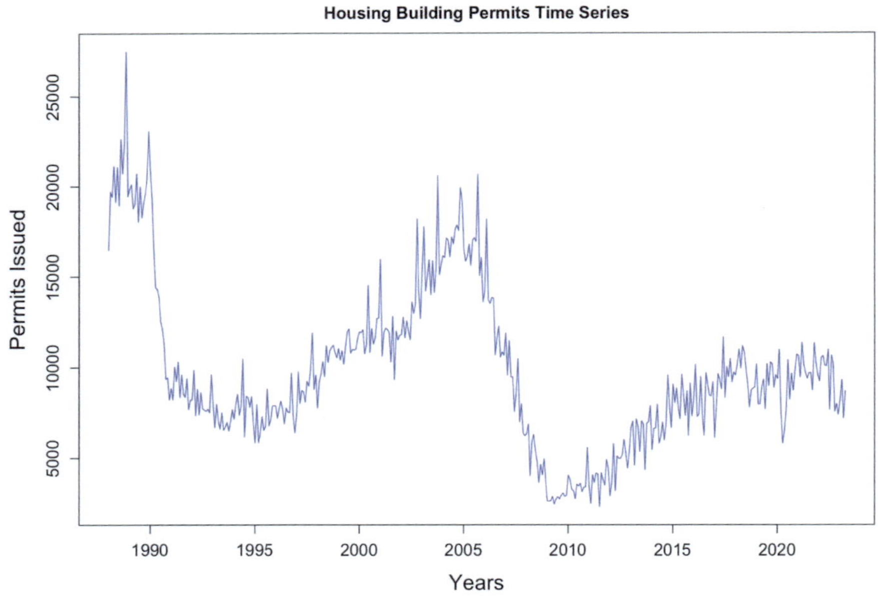

Fig. 6.1 Observed values of the housing permits time series

necessitate the relocation of entire communities, with profound implications for economic sustainability, livelihoods and community cohesion [2].

To explore these dynamics empirically, we analyze a dataset consisting of monthly housing permits issued in the state of California from 1988 to 2023. The dataset is publicly available through the Federal Reserve Bank of St. Louis's FRED platform, drawing from U.S. Census Bureau data on new private housing units authorized by building permits (U.S. Census Bureau [3]).

Figure 6.1 below presents the time series of monthly housing permits. It is evident that the series reflects a complex, non-linear structure, featuring cyclical peaks and troughs, short-term fluctuations, and longer-term shifts that likely correspond to macroeconomic conditions such as recessions, recoveries, and policy changes. These features necessitate a modeling approach capable of decomposing the series into its fundamental components: trend, seasonality, and random variation.

Accurate forecasting of future housing permit activity is of central importance to policymakers, planners, and researchers concerned with climate adaptation and urban development. The ability to parse these components effectively informs our understanding of not only historical patterns but also the potential trajectory of housing development under various future scenarios.

We use the FRED's CABPPRISA series at the state level for pedagogical reasons (and convenience, as well). We are aware that a real-world analysis of a proposed managed retreat policy would require data at a much higher level of geographic granularity. Data at the state level will mask the idiosyncratic variation that occurs at smaller scales. This is especially true for natural, social, and economic processes affected by climate change, and the related adaptation actions. We trust that the analysts following the proposed workflows will incorporate all the necessary levels of detail in their work.

6.3 Data Preparation

We use the package `readr` to facilitate the reading of different types of data files. The package `forecast` provides methods and tools for displaying and analyzing univariate time series forecasts. Notice that your `csv` file may be located in a different folder. After that, we use the function `ts` (from the base package `stats`) to transform the data (originally stored as a data frame) into a time series object. The R environment uses objects of the time series class (ts) to address the specificity of such data structures. Next, we plot the observed data set using the command `plot`.

```
# Reads the data file
Calif_Housing_Permits <- read_csv("./data/Calif_Housing_Starting.csv")
# This dataset is available from this book's website.
permits_TS <- ts(Calif_Housing_Permits[[2]], start=c(1988,1), end=c(2023,4),
     frequency=12)
#Transform dataframe into a time series
```

We prepare the housing permits dataset for time series modeling and forecasting. Visualization and appropriate partitioning of the dataset are essential preliminary steps before model estimation, enabling us to assess data structure and to validate model performance objectively.

```
# set graphing parameters
stats::plot.ts(permits_TS,
           main = "Housing Building Permits Time Series", ylab = "Permits
                Issued",
           xlab = "Years", col = "royalblue" )
```

We begin by visualizing the time series of monthly building permits using the `plot()` function from the `forecast` package in R (Fig. 6.1). While this function provides a clear and simple representation of the data, readers are encouraged to explore other visualization tools, such as those offered by the `TSstudio` or `ggplot2` packages, for interactive or web-based outputs. Functions such as `TSstudio::ts_plot()` enable dynamic visualization, which can be particularly useful in the context of building interactive dashboards for supporting decision-making.

Dataset Partitioning: Training and Testing Sets

In accordance with best practices in time series forecasting [1], we partition the dataset into two disjoint subsets:

- A **training set**, which will be used to estimate the parameters of the forecasting models.
- A **testing set**, which will serve as a hold-out sample for evaluating out-of-sample predictive performance.

Given the temporal nature of the data, random sampling is inappropriate. Instead, we follow a chronological split, allocating the first 74% of the data for training and the remaining 26% for testing (which is within the recommended range of 70–80%, and 20–30%, respectively, for each set [1]). Specifically, in this example, they are as follows:

- The training dataset contains 336 monthly observations, spanning January 1988 to December 2015.

- The testing dataset comprises 88 monthly observations, from January 2016 to April 2023.

```
# Training and Test series
train_ts <- window(permits_TS, start=c(1988,1), end=c(2015,12))
test_ts <- window(permits_TS, start=c(2016,1), end=c(2023,4))
```

If it is of interest to the reader, we suggest exploring the `timeTK` and `TSstudio` packages as they have specific functions to split the dataset into training and testing sets using parameters that refer to fixed percentages, instead of the reference to starting and ending points we used in the code above.

This chronological split enables us to compare model forecasts against observed values in the test set, thereby providing an empirical basis for evaluating forecast accuracy through standard metrics such as the Mean Absolute Error (MAE), Root Mean Squared Error (RMSE), and Mean Absolute Percentage Error (MAPE). The concept and application of the Mean Absolute Scaled Error (MASE) are discussed in Hyndman and Koehler [4].

Such error metrics will be discussed in subsequent sections on model evaluation. They provide crucial insights into the model's ability to generalize beyond the training data—an important criterion in high-stakes policy applications such as climate-sensitive land-use planning.

The code below plots the training and testing datasets, and it is shown in Fig. 6.2.

```
# set graphing parameters
par(cex.axis=0.8, cex.lab=0.8, cex=0.8)
plot(train_ts, col="royalblue", main="", ylab="Permits Issued", xlab="Years",
    xlim=c(1988,2023))
lines(test_ts, col="indianred")
legend("topright", lty=1.5, col=c("royalblue","indianred"),
      legend=c("Training set","Testing set"),bty="n")
```

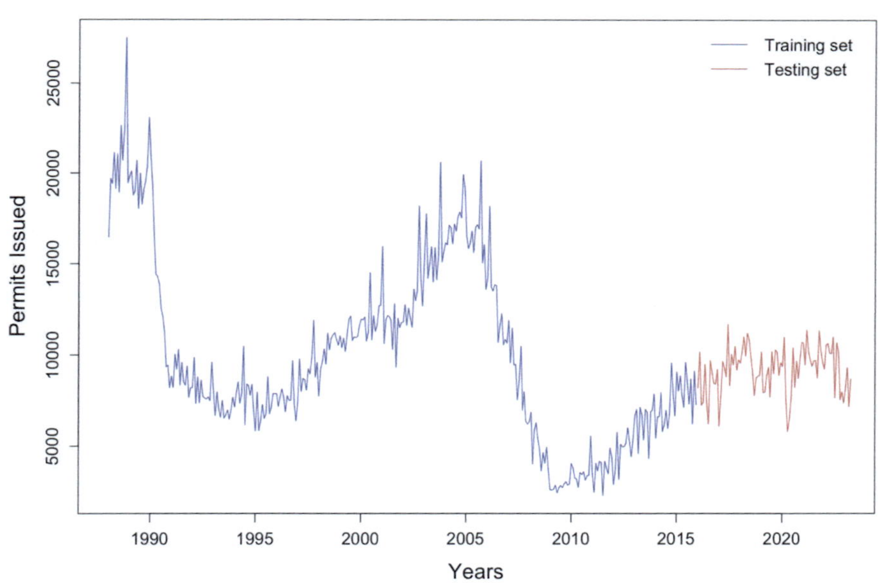

Fig. 6.2 Training and testing sets

6.4 Characterizing Time Series

Time series analysis can be a valuable tool for understanding complex problems, but it is important to be aware of its limitations. To that effect, we must first understand the main components of a series. Extracting those elements is commonly known as *characterization.*

We can create a plot of the observed data, trend, seasonal, and random components with the `decompose()` and `autoplot()` functions. The first one is from the base package `stats`, while the second one is from the `forecast` package. One advantage of this approach is the print-ready graphic quality. Figure 6.3 shows the decomposition of the housing permits dataset.

```
# set graphing parameters
par(cex.axis=0.6, cex.lab=0.6, cex=0.8)
# Plots components of the time series object
permits_TS |>
  decompose() |>
  autoplot()
```

Let us consider some details of the time series components plotted above. The main difference between cyclical and seasonal fluctuations is based on whether the temporal sequence obeys a fixed frequency or not; if the frequency changes over time, it is *cyclic,* and if it is fixed (daily, weekly, monthly, or related to some characteristic

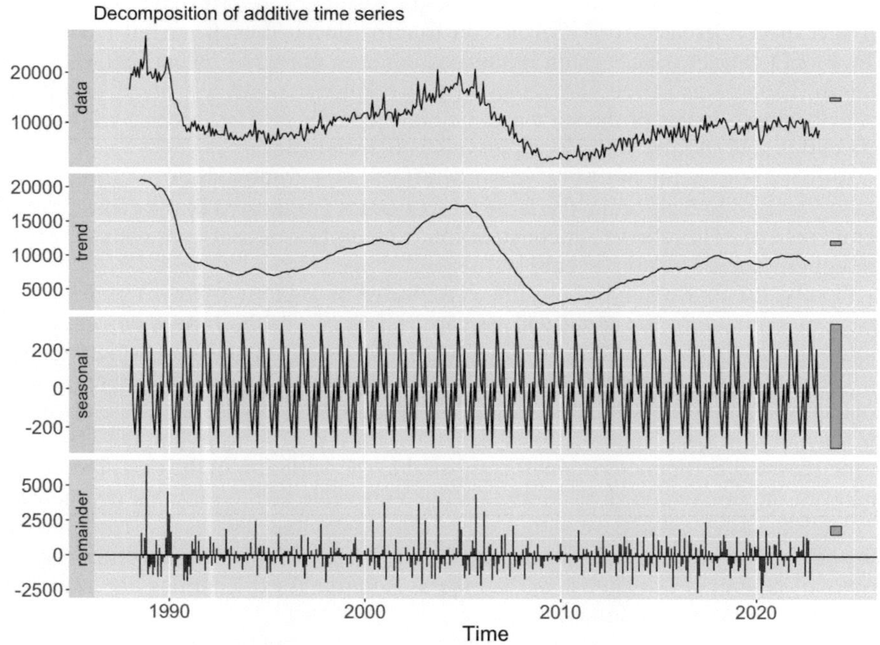

Fig. 6.3 Decomposition of the housing permits time series

of the calendar), the changes are *seasonal*. For a fuller discussion, see Hyndman and Athanasopoulos [1]; they are also the authors of the powerful `forecast` package [5].

The top panel in the decomposition plot shows the **observed data**, in which we can see all short-term fluctuations and long-term trends. The second panel shows the overall **trend** observed in the data. You can see that the short-term fluctuations have been removed, and only longer waves are present. The most common process to reduce those short-term variations is to compute a moving average. The longer the period over which the average is computed, the more short-term effects will be removed. For instance, a 6-month moving average will show a flatter pattern, with fewer oscillations than one based on a 3-month cycle.

The next panel (third block) shows the **seasonal effects**. It is rather simple to identify the seasonal fluctuation that may follow, for example, the summer-winter sequence as the issuing of building permits declines in autumn and increases again in spring. There are however, other minor cycles that may not be easily identifiable. Those may be the outcome of the built-in algorithm that is picking up other monthly (or multi-month) changes. The package computes a series of *seasonal coefficients* to capture the magnitude of the effects; they are later used in forecasting future values with the appropriate seasonal cycles.

Although this is not a book on the economics of the housing market in California, it is necessary to comment on a rather unusual feature of the dataset. Traditionally, housing permit data follow a similar seasonal pattern to the construction seasons, but are shifted a few months earlier. Thus, we would observe an increase in housing permits in winter, in anticipation of the construction uptick in spring and summer, and would expect a decline in early fall, as construction slows down in late fall and winter. The FRED dataset we are using in this section does not quite reflect that expected pattern. The seasonal effects are rather weak in relation to other, longer cycles related to the business cycles of the state and national economies. The forecast values will reflect this and, in turn, show weak seasonal characteristics.

Another reason for the weak seasonal effects is a combination of climate and demographics. In California, the largest urban areas are located in the coastal, temperate regions. This allows for construction to take place year-round, with very few disruptions due to weather events.

Finally, the fourth panel shows the **remainder** or random effects. These are the result of subtracting the trend and seasonal effects from the base data. They are interpreted as fluctuations that occur randomly, and we cannot easily explain them.

The grey bars to the right of each panel show the relative scales of the components. Each grey bar represents the same length, but because the plots are on different scales, the bars vary in size. The large grey bar in the third panel shows that the variation in the seasonal component is small compared to the variation in the data, the trend, and even the remainder. If we were to reduce the scale of the seasonal component to match the other panels, we would see an almost flat line. This is also captured by the scale of the vertical axis in each panel. The scale of the seasonal component is orders of magnitude smaller than the others. Additional explanations are available in Hyndman and Athanasopoulos [1].

You can see that the seasonal component is simply a repetition of a single pattern for each of the ten years in the dataset. These are the coefficients statistically estimated by the `decompose` function. We could extract those coefficients and turn them into a vector; then we could subtract that vector from the original data, and the result would be a *deseasonalized* time series. That is, a new dataset without the seasonal effect. That would allow us to compare monthly permits, regardless of whether it is summer or winter, and determine whether the number of permits was higher or lower between the two dates, without the relative effects of seasons.

6.5 Brief Introduction to ARIMA Models

Time series analysis is often criticized for neglecting the true causes of a phenomenon in favor of analyzing patterns of variation over time. This criticism is valid, as failing to understand the true causes of a problem can make it more difficult to solve, or impossible to determine if one has arrived at the correct solution. However, even without a full specification of the underlying causes, time series analysis can be a valuable tool for understanding long-term trends and as the starting point for developing more sophisticated analytical tools.

Time series is a vast data domain, and there are different ways to classify its components. One distinction is between *deterministic* and *stochastic* series. In a deterministic series, each observation can be calculated exactly as a function of one or more previous observations. In a stochastic series, each observation is a realization of a random variable. Stochastic series can be further classified into *stationary* and *non-stationary* series. A stationary series has a constant mean and variance over time, while a non-stationary series does not (Peng [6], Hyndman and Athanasopoulos [1]).

Forecasting techniques are, for the most part, available for stationary series; therefore, ensuring that a series is stationary is a prerequisite for forecasting. A non-stationary series can be transformed into a stationary series by differencing or taking the natural logarithms. A **difference-stationary** series is a non-stationary series that has been transformed by *differencing* (that is, subtracting consecutive observations). A **trend-stationary** series is a non-stationary series that has been transformed by taking the natural logarithm.

Let us consider the main components of the ARIMA models, and the three different ways in which we can estimate the value of a given observation in the series. First, an observation can be the result of averaging itself and one (or more) preceding observation(s). This is a moving average process or *MA(q)*. For instance, the symbol MA(q = 2), or simply MA(2), indicates that each observation could be understood as the average between itself and two prior observations. The process yields a 'moving average coefficient' that can be used to estimate future values.

Second, an observation can be estimated as a function of one or more previous observations. This is an auto-regressive process or *AR(p)*. For instance, the symbol AR(2) indicates that each observation is a linear combination of the two previous values. This process is analogous to a regression model, and produces the corresponding auto-regressive coefficients that can later be used to estimate future values.

Finally, we could consider an integration term of order d or *I(d)*. The function of this term is to transform the series into a stationary one by differencing it. The parameter *d* determines the order of the differentiation process. *Differencing* simply means calculating the difference between two consecutive observations.

When considered together, these terms are the foundation of the ARIMA process, and it is specified as ARIMA (p, q, d). This is also known as the Box-Jenkins method.

To complement and extend the discussion above, see the excellent work of Box et al. [7] and Chatfield and Xing [8].

6.6 Forecasting with Built-in Functions

In the characterization process we just completed, we have identified the series' overall trend and its seasonal components. Since the observed data exhibits more complex patterns, we also determined a remainder or random component (one that cannot be explained by the trend, the seasonal effects, and the combination of the two). We will proceed in two steps. First, we will use the built-in functions in the `forecast` package to create a first approximation to our final forecast. Then, we will use the manual and automated options to build ARIMA models to estimate the future values of the series.

Using the built-in functions, we can model four main processes. First, the *Mean* approach, in which every future value is estimated as the average of all previous observations. Second, the *Random Walk (naïve)* (or simply, naïve) method, in which the future value is estimated as the last observed value. Third, the *Random Walk (with drift)* (or simply, drift) method, in which the future value is estimated as the last observed value plus the average of the differences between consecutive observations. Finally, the *Seasonal Naïve* method, in which the future value is estimated as the last observed value from the same season. For different datasets, seasons will refer to different units of time, such as the same day of the week, or the same month of the year, or other special types, as the holiday season.

The plot below shows the training series with the different forecasting methods. The red line represents the *Mean* method, the blue line the *Naïve* method, the orange line the *Drift* method, and the green line the *Seasonal Naïve* method. We compute the estimates directly in the code that builds the plot in Fig. 6.4. This is very efficient as long as one does not need those estimates for another purpose, as they are not

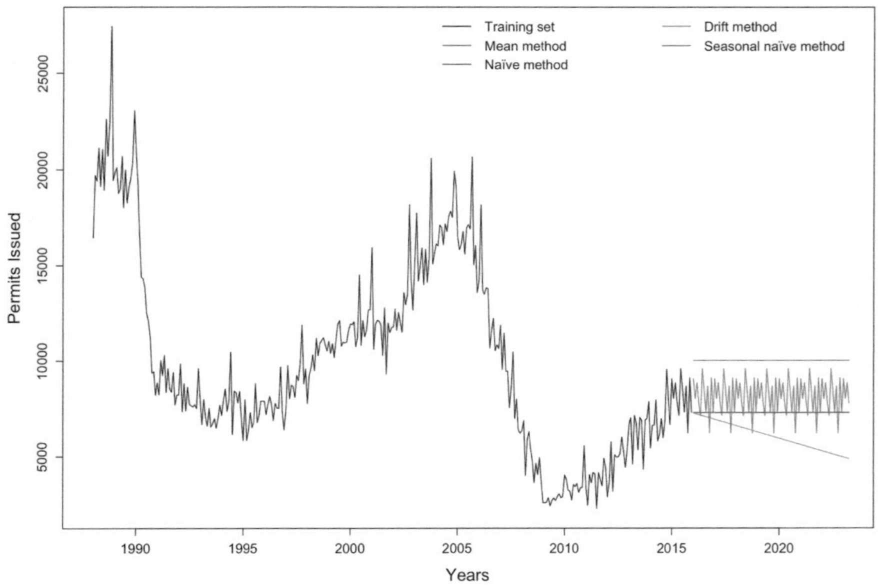

Fig. 6.4 Training data and forecast values using built-in methods

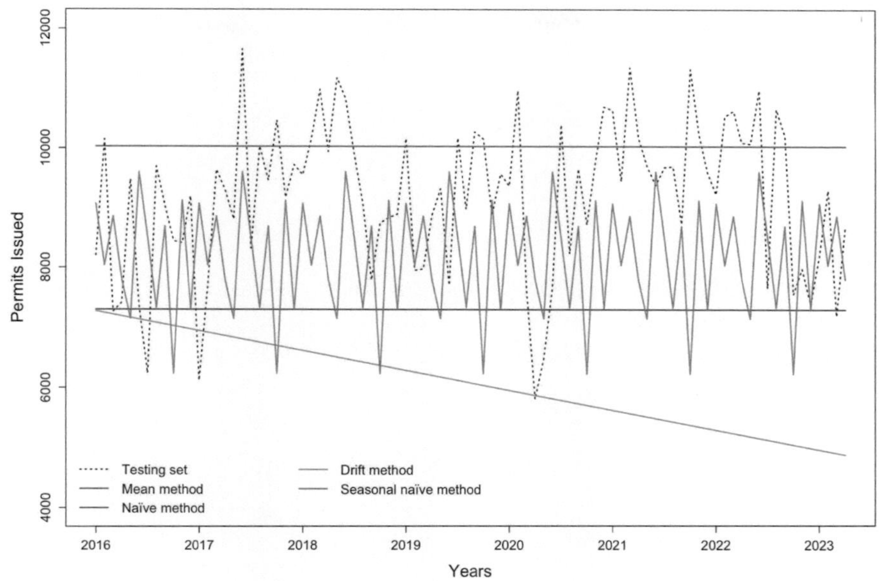

Fig. 6.5 Testing data and forecast values using built-in methods

saved in the system. If they were to be used elsewhere, we need to extract them by assigning them to separate objects.

For the results plotted in Figs. 6.4 and 6.5, and listed in Table 6.1, we have used the built-in function `meanf`, `rwf` (with and without `drift=TRUE`), and `snaive` to build the corresponding models.

```
plot(train_ts, main="", ylab="Permits Issued",
    xlab="Years", xlim=c(1988,2023))
# plot forecasting for 5 years according to four methods
lines(meanf(train_ts,h=88)$mean, col="red")
lines(rwf(train_ts,h=88)$mean, col="blue")
lines(rwf(train_ts,drift=TRUE,h=88)$mean, col="orange3")
lines(snaive(train_ts,h=88)$mean, col="green4")
# legend
legend("topright", lwd=1.25, col=c("black","red", "blue",
                        "orange3", "green4"),
      legend=c("Training set","Mean method","Naïve method",
            "Drift method", "Seasonal naïve method"),bty="n", ncol=2)
```

Let us now zoom in to see the relationship between the testing data and each of the forecasting methods.

```
# plot the test set
dates <- seq(as.Date("2016-01-01"), by = "year", length = 88)
plot(test_ts, main="", ylab="Permits Issued", xlab="Years",
    xaxt = "n", lty=3, lwd=1.5, ylim = c(4000,12000))
axis(1, at = lubridate::decimal_date(dates), labels = format(dates,"%Y"))
# plot forecasting for 4 years according to four methods
lines(meanf(train_ts, h=88)$mean, col="red", lwd=1.5)
lines(rwf(train_ts, h=88)$mean, col="blue", lwd=1.5)
lines(rwf(train_ts, drift=TRUE,h=88)$mean, col="orange3", lwd=1.5)
lines(snaive(train_ts, h=88)$mean, col="green4", lwd=1.5)
```

Table 6.1 Accuracy meansures of different built-in models

	ME	RMSE	MAE	MPE	MAPE	MASE	ACF1	Theil's U
Mean method								
Training set	0.00	4916.35	3959.01	–31.80	55.39	1.88	0.94	
Test set	–838.74	1520.52	1164.66	–11.48	14.46	0.55	0.30	1.07
Random Walk (naïve) method								
Training set	–27.37	1662.57	1213.13	–1.95	13.84	0.58	-0.47	
Test set	1893.19	2278.75	1999.87	18.88	20.62	0.95	0.30	1.44
Random Walk (w/drift) method								
Training set	0.00	1662.35	1214.68	–1.59	13.83	0.58	–0.47	
Test set	3111.03	3465.73	3151.00	32.26	32.91	1.50	0.51	2.18
Seasonal naïve method								
Training set	–466.00	2942.81	2107.04	–8.97	26.31	1.00	0.68	
Test set	1036.85	1928.21	1637.30	9.25	17.66	0.78	–0.05	1.24

```
# legend
legend("bottomleft", lwd=1.5, lty=c(3,1,1,1,1),
      col=c("gray4","red", "blue", "orange3", "green4"),
      legend=c("Testing set","Mean method","Naïve method",
             "Drift method", "Seasonal naïve method"),bty="n", ncol=2)
```

Accuracy of built-in methods

In many cases, we can examine the accuracy of the forecasting models directly from the plots we have built. Although this is convenient, it is not the most efficient way to compare the models. The `forecast` package provides a function called `accuracy` that calculates the most common error measures for each model. The code below calculates the accuracy of the four models we have used. Table 6.1 shows the results.

```
# Define common parameters

h <- 88

# Create list of forecasting methods and their corresponding functions
forecast_methods <- list(
  "Mean" = function(ts) meanf(ts, h = h),
  "Naive" = function(ts) rwf(ts, h = h),
  "Drift" = function(ts) rwf(ts, drift = TRUE, h = h),
  "Seasonal Naive" = function(ts) snaive(ts, h = h)
)

# Generate forecasts and calculate accuracy in one step
accuracy_results <- do.call(rbind,
      lapply(forecast_methods, function(method) {
      forecast_obj <- method(train_ts)
      forecast::accuracy(forecast_obj, test_ts)})
)
```

```
knitr::kable(accuracy_results, digits = 2,
         caption = "Accuracy Meansures of Different Built-in Models") |>
  kableExtra::group_rows(group_label = "Mean method",
                   start_row = 1, end_row = 2) |>
  kableExtra::group_rows(group_label = "Random Walk (naïve) method",
                   start_row = 3, end_row = 4) |>
  kableExtra::group_rows(group_label = "Random Walk (w/drift) method",
                   start_row = 5, end_row = 6) |>
  kableExtra::group_rows(group_label = "Seasonal naïve method",
                   start_row = 7, end_row = 8) |>
  kable_styling(latex_options = c("HOLD_position","scale_down"))
```

The list of acronyms in the table above and their corresponding definitions are listed below.

`ME`: Mean Error
`RMSE`: Root Mean Squared Error
`MAE`: Mean Absolute Error
`MPE`: Mean Percentage Error
`MAPE`: Mean Absolute Percentage Error
`MASE`: Mean Absolute Scaled Error
`ACF1`: Autocorrelation of errors at lag 1
`Theil's U`: Comparison to a random walk (naïve) model

6.7 Manual ARIMA Forecasting

We will now use the ARIMA model to forecast the future values of the series. As we discussed above, an ARIMA model has three parameters, the autoregressive component (AR), the integration component (I), and the moving average component (MA). The AR component is the number of lags used to estimate the current value, the MA component is the number of moving averages used to estimate the current value, and the I component is the number of differences used to transform the series into a stationary one. The code below uses an ARIMA(1,0,1) model to forecast the 40 months that follow the training dataset. The selection of parameters (1, 0, 1) was somewhat arbitrary since, at this time, we do not have a clear insight into the optimal set of parameters. The parameters are AR(1), I(0), and MA(1); that means that every observation will be estimated as a combination of a linear function of the previous observation and the average of itself and the previous observation. The parameter I(0) indicates that we are not differencing the series.

6.8 Automated ARIMA Forecasting

We indicated above that we used the ARIMA(1, 0, 1) model specification as a way of getting a first approach to the optimal parameters. Fortunately, we can automate this process by using the `auto.arima` function (from the forecast package). This function estimates the optimal combination of components (auto-correlation, integration, and moving average) that fits the observed data the best. The model generated by the automated function has an ARIMA(1, 1, 2) (2, 0, 2) [12] formulation.[1] The `auto.arima` function correctly identifies the strong seasonal component. Consequently, it added a set of *seasonal parameters*, in this case, the second set of three values (2, 0, 2) to capture the seasonal effects.

6.9 Other Methods

The last line in each block of code uses the `HoltWinters` function (from the base package `stats`) to fit a model that we later use to forecast the series into the future. The Holt-Winters model is an exponential smoothing approach, and its parameters can be modified to match different modeling conditions. For example, the `alpha` parameter adjusts the effect of more recent values, with higher values weighting more recent observations more heavily. Similarly, higher values of `beta` (or trend effects) assign higher power to recent trend slopes. The `gamma` parameter modifies the seasonal adjustment component. If `gamma = FALSE`, the model will not adjust for seasonal effects. We invite the reader to experiment with different values of `alpha`, `beta`, and `gamma` to understand how they affect the forecast. We opted for using the model's default parameters.

The following code block builds the models and calculates the forecast values for all three models, ARIMA, auto-ARIMA and Holt-Winters.

```
train.ARIMA <- forecast(arima(train_ts, order=c(1,0,1)),88)
train.autoARIMA <- forecast(auto.arima(train_ts),88)
train.HW <- forecast(HoltWinters(train_ts), 88)

fore.ARIMA <- train.ARIMA$mean
fore.autoARIMA <- train.autoARIMA$mean
fore.HW <- train.HW$mean
```

The plot in Fig. 6.6 shows the observed values of the series and the forecast values using the ARIMA, Auto-ARIMA, and Holt-Winters methods calculated in the code block above.

```
plot(train_ts, xlab ="Year", ylab = "Observed and forecast values",
    xlim=c(1988,2023))
lines(fore.ARIMA, col="green4", lwd=1.5)
lines(fore.autoARIMA, col= "orange3", lwd=1.5)
lines(fore.HW, col="blue", lwd=1.5)
legend("topright", lwd=2.5, col=c("green4", "orange3", "blue"),
     legend=c("ARIMA","Auto-ARIMA","Holt-Winters"),bty="n")
```

6.10 Complementary Plots

Most time series charts can include confidence intervals around the forecast values. The code below creates a plot with the observed values, the forecast values, and the confidence intervals for the ARIMA model. Similar plots can be created for

[1] The added element *[12]* refers to the periodicity of the seasonal effects. In this case, 12 months.

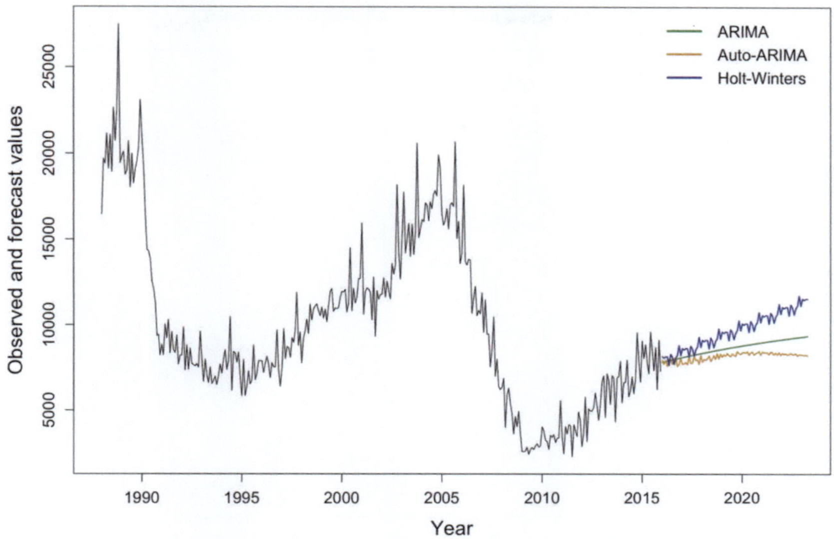

Fig. 6.6 Forecast with ARIMA, Auto-ARIMA and Holt-Winters methods

the autoARIMA and Holt-Winters models. We can see in the chart below that the confidence intervals are wider for the forecast values that are further in the future. This is a common feature of time series forecasting, as the uncertainty increases with the distance from the last observed value.

Without using too much math, we can show that errors widen as we move further into the future, meaning that our certainty declines as we move further away from the present. To that effect, we will consider the following thought experiment. For each point we forecast into the future, we will build a confidence interval to capture our level of certainty. That interval will have an upper and a lower limit; that is, the maximum and minimum value the estimate could have. Assuming that the upper limit is the value that ends up occurring, we will have to build another confidence interval around that upper limit. As you can see, for each iteration, the upper limit of the estimates keeps moving upward, possibly at a faster rate. The same applies to the lower limit of the confidence intervals. This shows that, if we connect all the upper and lower limits, we will have a widening range of values as we move further into the future. Figure 6.7 shows that for two default levels of confidence (the narrower one for 80% and the wider one for 95% confidence).

```
plot(train.ARIMA, showgap=FALSE,
    xlab ="Year",
    ylab = "Observed and forecast values")
```

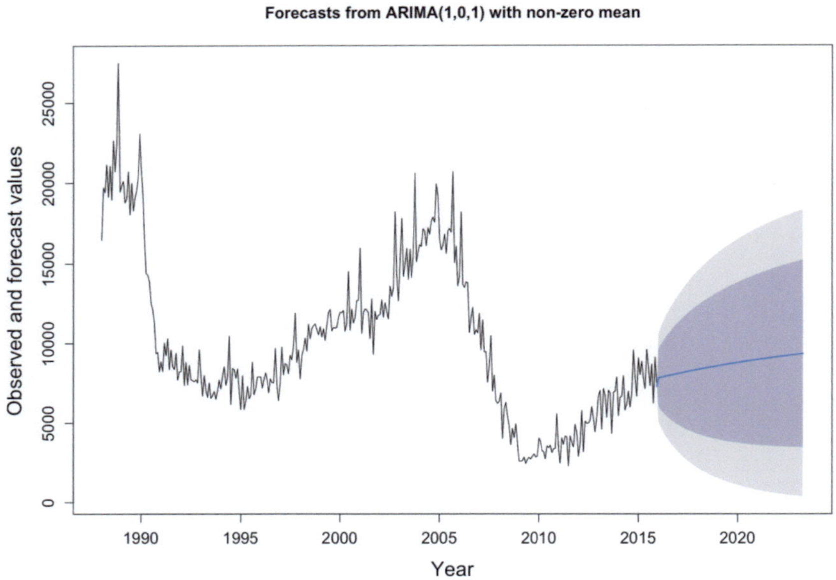

Fig. 6.7 Forecast with autoARIMA, showing confidence interval

References

1. Hyndman, Rob J., and George Athanasopoulos. 2018. *Forecasting: Principles and Practice*. English, 2nd edn. Australia: OTexts.
2. Hino, Miyuki, Christopher B. Field, and Katharine J. Mach. 2017. Managed retreat as a response to natural hazard risk. *Nature Climate Change* 7(5): 364–370. https://doi.org/10.1038/nclimate3252. https://doi.org/10.1038/nclimate3252.
3. U.S. Census Bureau. 2025. *New Private Housing Units Authorized by Building Permits for California [CABPPRIVSA]*. Federal Reserve Bank of St. Louis. https://fred.stlouisfed.org/series/CABPPRIVSA.
4. Hyndman, Rob J., and Anne B. Koehler. 2006. Another look at measures of forecast accuracy. *International Journal of Forecasting* 22(4): 679–688. https://doi.org/10.1016/j.ijforecast.2006.03.001. https://www.sciencedirect.com/science/article/pii/S0169207006000239. ISSN: 0169-2070.
5. Hyndman, Rob J., and Yeasmin Khandakar. 2008. Automatic time series forecasting: The forecast package for R. *Journal of Statistical Software* 27(3): 1–22. https://doi.org/10.18637/jss.v027.i03.
6. Peng, Roger. 2020. *A Very Short Course on Time Series Analysis*. https://bookdown.org/rdpeng/timeseriesbook/.
7. Box, George E. P., et al. 2015. *Time Series Analysis: Forecasting and Control*, 5th edn. Wiley.
8. Chatfield, Christopher, and Haipeng Xing. 2019. *The Analysis of Time Series: An Introduction with R*, 7th edn. Chapman & Hall/CRC Texts in Statistical Science Series. CRC Press, Taylor and Francis Group. ISBN: 978-1-351-25944-6.

Chapter 7
Bayesian Models

7.1 Introduction

At the heart of statistical analysis lie two distinct philosophical paradigms: the **frequentist** and **Bayesian** approaches. While both aim to draw inferences from data, their fundamental interpretations of probability and uncertainty diverge significantly, resulting in distinct methodologies for parameter estimation and hypothesis testing.

The *frequentist* perspective, often referred to as classical statistics, conceptualizes probability as the long-run relative frequency of an event. Within this framework, an unknown parameter of a population is regarded as an unobservable constant. Statistical inference, therefore, revolves around understanding what would transpire if an experiment were to be hypothetically replicated numerous times. For instance, in hypothesis testing, a frequentist assesses the probability of observing data as extreme as, or more extreme than, the current data, assuming the null hypothesis to be true. This value, the *p-value*, guides the decision to reject or fail to reject the null hypothesis at a pre-specified significance level. Similarly, a confidence interval is constructed to delineate a range of values such that, over repeated sampling, a certain percentage of these intervals would encapsulate the true parameter. It is relevant to note that a frequentist confidence interval does not express the probability that the true parameter lies within that specific interval. This approach generally necessitates larger sample sizes to yield robust conclusions and is often characterized by its perceived objectivity, as it relies primarily on observed data and predefined statistical procedures [1].

Conversely, the *Bayesian* approach defines probability as a degree of belief or a quantifiable measure of uncertainty concerning an event or an unknown parameter. Here, unknown parameters are treated as random variables and are consequently assigned probability distributions. The cornerstone of Bayesian inference is Bayes' Theorem, a mathematical framework for updating prior beliefs in light of new empirical evidence. This process begins with a prior probability distribution, which mathematically encapsulates existing knowledge or assumptions about the parameters before any data collection occurs. As new data become available, this prior distribution is combined with the likelihood function (representing the probability of observing the data given different parameter values) to yield a posterior probability distribution. This posterior distribution represents the updated and refined beliefs

F. DePaolis, *Workflows for Climate Change Adaptation Analytics*,
https://doi.org/10.1007/978-3-032-14004-3_7

about the parameters, having assimilated the information from the observed data. Unlike frequentist confidence intervals, Bayesian credible intervals directly express the probability that the true parameter lies within a given range.

A distinct advantage of the Bayesian methodology is its ability to provide meaningful inferences even with smaller datasets, as it explicitly incorporates prior knowledge. However, the selection of the prior distribution introduces a subjective element into the analysis, a point often highlighted by critics, although Bayesians contend that some degree of subjectivity is inherent in all statistical modeling.

In essence, while frequentist methods focus on the properties of procedures under hypothetical repetitions of an experiment, treating parameters as fixed, the Bayesian paradigm employs probability to quantify uncertainty about parameters themselves, continually updating these beliefs as new data emerge.

The choice between these two powerful approaches ultimately depends on the specific research question, the nature of the available data, and the analyst's philosophical stance regarding the definition and interpretation of probability. In the end, this is more a philosophical distinction than an operational one, as Yuzheng Sun says (https://www.statsig.com/blog/bayesian-vs-frequentist-statistics), but one we need to be aware of in our professional practice.

For extensive details, see Faulkenberry [2] and Gelman et al. [3].

7.2 Why Do We Need a Bayesian Approach to Time Series?

As I always share with my students and junior colleagues, I prefer simpler methods to those that are more convoluted. There is, of course, nothing wrong with experimenting with more sophisticated methods that are constantly being developed. But the downside is that those require, in general, lengthier and more complex explanations that do not always help understand the problem at hand. The added risk is that our audiences may be suspicious of our less straightforward contributions. In the case of time series, as we discussed earlier, we are already making an extreme simplification by considering the evolution of the variable of interest a function of the passage of time alone. To make things worse, we have also said that our level of confidence declines as we predict values further into the future, making the errors monotonically larger. One way to improve the perception of the quality of our forecasts—if not their objective quality—is to share our understanding of the magnitude of the underlying uncertainty inherent in our analyses. Bayesian methods help in this regard by providing a full posterior distribution of the model parameters.

Bayesian models also allow for the incorporation of prior knowledge or beliefs into the analysis, which can be particularly useful when data is scarce or noisy. They also provide a more intuitive interpretation of results, as they produce probability distributions for parameters rather than point estimates. This can help to quantify uncertainty and make more informed decisions based on the analysis. Many R packages provide a wealth of functions to implement these concepts for time series analysis. We will cover a small subset of those. For a comprehensive discussion of Bayesian statistics, see Gelman et al. [3] and McElreath [4].

7.3 The Buoy dataset

The data come from Station 46042 (LLNR 297) - MONTEREY, located 27NM WNW of Monterey, CA (36°47'5" N 122°23'46" W). The station is owned and maintained by the National Data Buoy Center. You can create your own dataset by downloading the data from the National Data Buoy Center [5]. The historical records for standard meteorological data are available in separate files by year. We have compiled ten separate datasets for the years 2013 through 2022, and combined them in a single time series object. You can find them on this book's website and follow the procedure in the code blocks below.

7.4 Data Preparation

Although it is true of all analytical techniques, it is particularly true of time series analysis that the data preparation stage is crucial for the success of the analysis. In the code below, we will read the data files, clean them, and create a time series object that is suitable for further analysis.

```
# looks inside the 'data' directory
f_path <- "./data"
datalist <- lapply(list.files(path = f_path, pattern = "*.txt", full.names =
     TRUE),
               function(x)read.table(x, header=T))
#assuming the same header/columns for all files
buoy_datafr = do.call("rbind", datalist)
```

The block above reads the data files from the `data` directory and combines them into a single data frame.[1] The data files are assumed to have the same header, which is a common practice when dealing with time series data collected from a single source using APIs or other standard formats. In this case, the data files contain information about the water temperature (`WTMP`) and other meteorological variables collected by the buoy station. The files have the same structure, with columns for year (`YY`), month (`MM`), day (`DD`), hour (`hh`), minute (`mm`), and water temperature (`WTMP`), among others. The combined data frame contains all the data from the individual files, making it easier to work with the entire dataset at once.

The next block of code cleans the dataset, selecting the necessary columns and converting the date and time columns into a single date-time column. This is a crucial step for time series analysis, as it allows us to work with a single date-time object rather than separate year, month, day, hour, and minute columns. It also filters the dataset to include only the observations taken at 12:50 PM, to ensure uniformity and consistency across different dates, but without the effect of hourly fluctuations.

[1] Please notice that there might be some slight differences if you are using a Windows computer.

```
# Filter for 12:50PM
buoy_datafr2 <- buoy_datafr[buoy_datafr$hh=="12" & buoy_datafr$mm=="50",]

# Create FULL DATE
buoy_datafr2$full_date <- lubridate::ymd(paste(buoy_datafr2$YY,
                                               buoy_datafr2$MM,
                                               buoy_datafr2$DD, sep= ' '))
```

Here, we create a much simpler dataset with only the date and water temperature (`WTMP`) columns. We also remove any observations where the water temperature is missing (represented by 999.0). We also limit the dataset to have only 52 weeks per year. We summarize the data by week to create a single weekly reading of water temperature averages (over all available dates for that week, which varies across the years).

```
# Dataset with DATE and WTMP. Removes missing values
buoy_WTMP <- buoy_datafr2[,c(19,15)] |>
  filter(WTMP != 999.0)

# Create weekly averages, rounded to three decimals
buoy_WTMP$week <- week(buoy_WTMP$full_date)
buoy_WTMP$year <- year(buoy_WTMP$full_date)

# Simplifies the dataset by keeping only 52 weeks each year. It calculates the
     weekly averages
WTMP_weekly <- as.data.frame(buoy_WTMP |>
                         filter(week!=53)|>
                         group_by(week, year) |>
                         summarise_at(vars("WTMP"), ~round(mean(.),3)))
```

Finally, we sort the dataset by year and week, and create a time series object (`ts`) that is suitable for further analysis. The `ts` object is a standard way to represent time series data in R. In this case, we set the start date to January 1, 2013, and the frequency to 52, indicating that the data is collected weekly, with a maximum of 52 weeks per year. It is possible that by imposing this limit, we are suppressing some data corresponding to the latest days of December for each year. We do not consider this detrimental to understanding how the techniques work or to the results presented in the examples.

```
# Sort the dataset so it is suitable for time series transformation
WTMP_weekly <- WTMP_weekly|>
  ungroup() |>
  arrange(year, week)
# Create time series
WTMP_weekly_TS <- ts(WTMP_weekly$WTMP, start=c(2013,1,1), frequency = 52)
```

7.5 Understanding the Time Series

One of the most effective ways of understanding the data in a time series is by creating a *decomposition* of the data. But what exactly does it mean? The decomposition of a time series is the separation of the data into its different components. The component is the *trend*, that is, the overall behavior over the entire time period. This could show an upward, downward, combined, or null trend. The second component is the *seasonal* effect, or fluctuations that take place at regular intervals, such as the distinct periods of a year (summer, fall, spring, winter), or other regular occurrences. The final component is the *random* element, also known as *residual*. This is the remainder

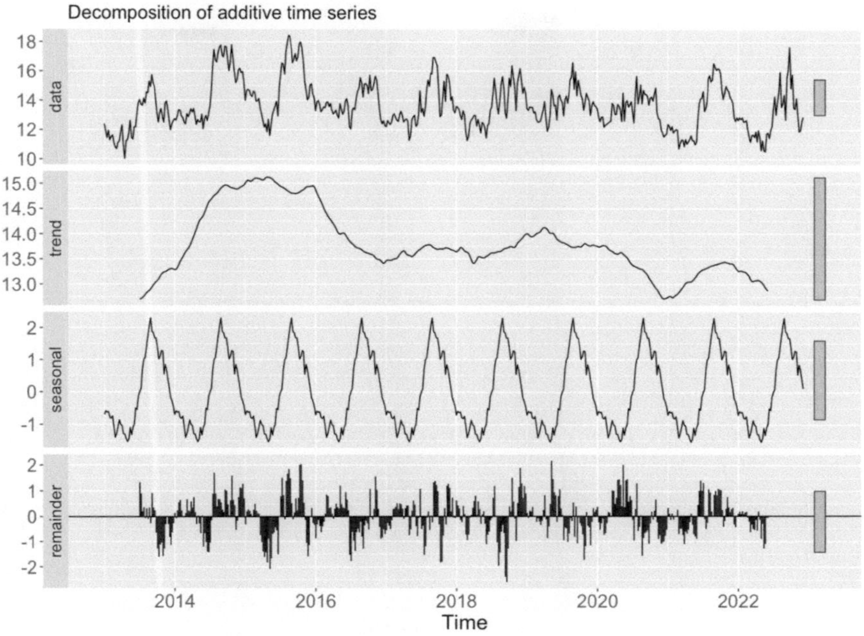

Fig. 7.1 Decomposition of the Buoy time series

after accounting for the trend and seasonal effects, and is usually unpredictable as it does not follow a discernible pattern. Several R packages include functions that create both the numeric and graphic decomposition of a time series. The code below creates the plot in Fig. 7.1 showing the decomposition of the water temperature data.

```
# set graphing parameters
par(cex.axis=0.8, cex.lab=0.8, cex=0.8)
# Plots components of the time series object
WTMP_weekly_TS |>
  decompose() |>
  autoplot()
```

In the *decomposition* plot, we can see the actual data, the long-term trend, the seasonal component, and the residual or remainder. Unlike the house permits dataset we used in the previous section, the ocean water temperature dataset has a very strong seasonal effect. This is somewhat expected as water temperature fluctuates at regular intervals during the year. We also observe that the remainder component is relatively small, with fluctuations within ±2 °C.

7.6 Training and Testing Sets

```
# Split into training and testing sets
train_WTMP <- window(WTMP_weekly_TS, start = c(2013, 1), end = c(2020, 52))
test_WTMP <- window(WTMP_weekly_TS, start = c(2021, 1), end = c(2022, 49))
```

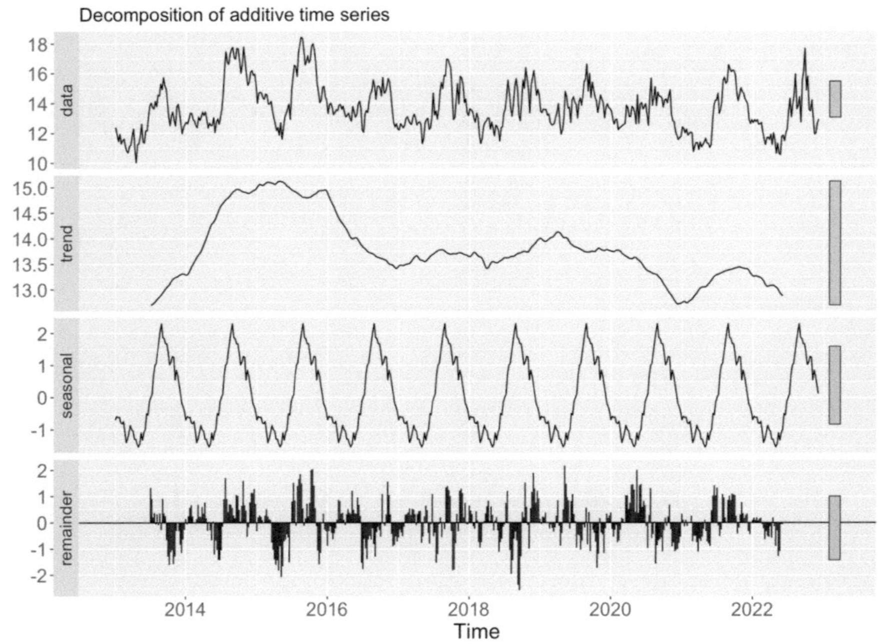

Fig. 7.2 Training and testing sets

Like we did in the section on ARIMA, here we split the dataset into a training and a testing set. This allows us to compute the parameters for all models using the training set. Then we can compare the forecast values and the testing set to assess how well the forecasting model fits data that have not been used to estimate the model parameters. Once we are satisfied with the performance of a particular model, we can use it to forecast values for an *out-of-sample* time period.

Figure 7.2 shows the training and testing split from the original water temperature dataset.

```
# set graphing parameters
par(cex.axis=1, cex.lab=1, cex=1)
plot(train_WTMP, col="royalblue", main="", ylab="Water temperature", xlab="Time
    ", xlim=c(2013,2023))
lines(test_WTMP, col="indianred")
legend("topright", lty=1.5, col=c("royalblue","indianred"),
     legend=c("Training set","Testing set"),bty="n")
```

7.7 Models

In this section, we use the `bayesforecast` package. It is one of the many R packages that implement Bayesian methods for time series analysis. The package is based on the `Stan` software, which is a probabilistic programming language for Bayesian inference. In recent years, there has been a significant expansion in the number of R packages that are Bayesian analysis-oriented. There are multiple reasons for such an abundance of options, including the increase in computing power, the growing interest of researchers and developers in techniques with applications in machine learning (ML), and the need to address the demands of business intelligence approaches.

For the best part of the last decade, there have been some interesting developments taking place in powerhouses like Google and Meta. For instance, Scott and Varian [6] (both economists at Google at the time of their paper) present a method for forecasting using an ensemble prediction averaging over different combinations of predictors. In their paper, they use the `bsts` package, which optimizes modeling options for fitting Bayesian structural time series (BSTS) models.[2] The `bsts` package is now part of the `bayesforecast` package, which we use in this section.

The `bayesforecast` package allows for the implementation of alternative models, such as ARIMA, Multiplicative Seasonal ARIMA, GARCH, ARMAX, Dynamic Harmonic Regression, and many others. We will cover only a few alternatives because some are not suitable for series with strong seasonal effects, such as the one used in this section. We encourage the reader to explore the package's documentation for further details.

7.7.1 Stan ARIMA

The first model we built is an ARIMA model using a Bayesian approach. As is often the case, we may not have a way to determine the best AR, I, and MA parameters. In this case, we use a (1, 0, 1) combination as the starting point. Because we already determined that there is a substantial seasonal effect in the data, we introduce a seasonal term in the model to capture those effects. In the code below, we use a (1, 1, 1) parameter to represent the inferred effects of seasonality.

```
WTMP_sarima <- stan_sarima(ts = train_WTMP, order = c(1,0,1),
           seasonal= c(1,1,1), prior_mu0 = student(mu=0, sd=1,df=7))
```

```
check_residuals(WTMP_sarima)
```

Residual plots

The built-in function `check_residuals` from the `bayesforecast` package provides a comprehensive view of a model's performance by analyzing its errors or residuals, that is, the difference between the actual and estimated values. The function produces a plot with five separate panels.

The top panel shows the value of the expected errors over the entire series. This gives you a better understanding of the range of errors, as well as whether there is an identifiable pattern, which would indicate a bias in the model. In Fig. 7.3 for the `Stan SARIMA` model, we can see that the errors fluctuate around $\pm 2\,^{\circ}$C. We can also see that the errors appear to be randomly distributed.

The next panel, on the left in the middle row, is a histogram of the errors. The corresponding normal distribution curve serves as a reference to determine how close

[2] The package `bsts` provides an alternative plot to show the decomposition of a time series. We recommend the reader explore it as well.

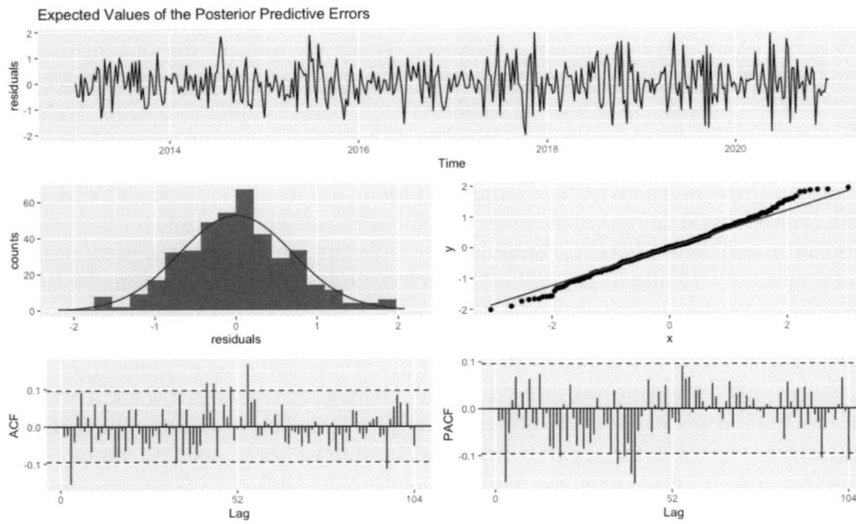

Fig. 7.3 Residuals Stan SARIMA model

the errors are distributed around zero. The panel on the right in the middle row is called *Normal Q-Q* plot, and it displays the errors' deviation from normality and other distributional attributes. Among those, skewness (if the points were to display an oddly curved pattern), kurtosis (in the case of an s-shaped pattern), or the presence of outliers (points that are located too far away from the reference line).

The bottom panels are the AutoCorrelation (ACF) and Partial AutoCorrelation (PACF) functions. The ACF is a measure of the correlation of the time series and a lagged version of itself. The PACF is also a measure of the correlation between the time series and itself, but after considering the effects of shorter lags. Both the ACF and PACF complement decomposition plots in understanding the presence and magnitude of seasonal effects, and prevailing trends (Hyndman and Athanasopoulos [7] and Peng [8]). In the subsequent sections, we include a residual analysis after building each model.

7.7.2 *Auto SARIMA*

We can also generate SARIMA models automatically. This function returns the best ARIMA model based on the value of the BIC (Bayesian Information Criterion). One of the advantages of using the BIC criterion is that it prioritizes simpler, lower-order models, which are, in general, quite efficient and relatively easy to explain. Figure 7.4 shows the residual analysis of the AutoSARIMA model.

```
# auto SARIMA
WTMP_auto <- auto.sarima(train_WTMP, seasonal = TRUE, chains=4, iter=1000,
    stepwise = TRUE)
```

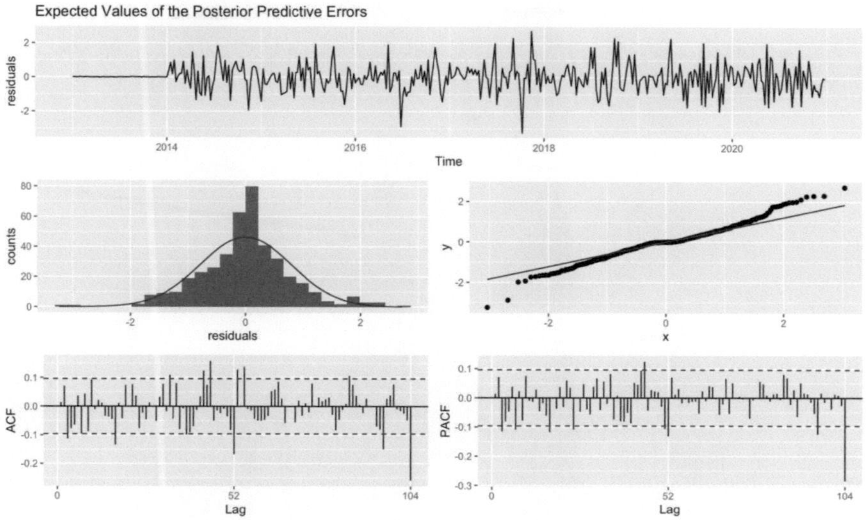

Fig. 7.4 Residual analysis of Auto SARIMA model

From the summary of the model, we see that the formulation that was derived from the automated process is equivalent to an ARIMA (2, 1, 1) (1, 1, 0) process.

```
check_residuals(WTMP_auto)
```

7.7.3 Additive Linear State Space Model-SSM

The most salient characteristic of the SSM models is that there is a *time-varying* relationship between consecutive observations. We do not commonly know that underlying relationship (sometimes called a *state*), but we do observe the changes in the values over time. These models are sometimes called BSTS (Bayesian structural time series) or Kalman filter models [8]. They are now implemented in a variety of languages (most noticeably, Python) as well as development environments, such as TensorFlow [9], and have wide application in machine learning. The `bayesforecast` package implements *SSM* models with different parameters to account for attributes of the time series, such as seasonality, effect of adjacent observations, global and local trends, periodicity, and number of iterations to run, among others.

Although we are not covering it in detail, it is important to consider the work by Brodersen et al. [10] at Google. They developed the R package `CausalImpact` to implement their approach to inferring causal impacts in econometric models based on Bayesian time series. This can be very effective in assessing the impact that policy interventions may have on outcomes over time.

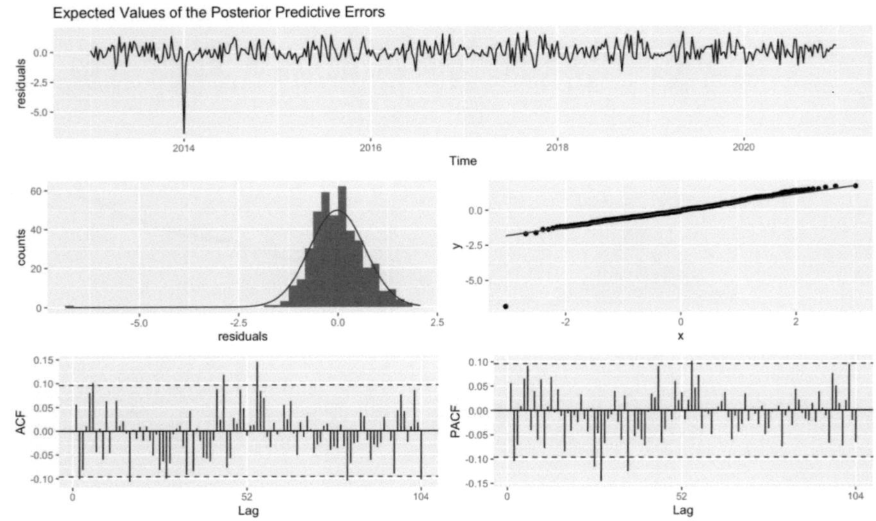

Fig. 7.5 Residual analysis of State Space model

```
# SSM - Additive linear State space model.
WTMP_ssm <- stan_ssm(train_WTMP, trend = TRUE, damped = TRUE,
                     seasonal = TRUE, stepwise = TRUE, iter = 1000)
```

```
check_residuals(WTMP_ssm)
```

The residual plot in Fig. 7.5 for the SSM model shows some peculiar attributes. For example, we see a' disproportionately large error around the beginning of 2014 (deep spike well below 5 °C). That event also appears in the two middle panels. The bottom panels showing the ACF and PACF, however, show very tenuous effects of lags.

7.7.4 *Random Walk—Naïve*

A random walk model is one in which the next observation is estimated as the current observation plus or minus a random error. It is equivalent to ARIMA (0, 1, 0) model. In the case below, we set the seasonal parameter to *TRUE*, and this creates a model equivalent to an ARIMA (0, 0, 0) (0, 1, 0), in which the values are estimated without regard to previous (lagged) observation and only the seasonal term is *differenced* once.

```
# naive -#- random walk equivalent to ARIMA(0,0,0)(0,1,0)m for seasonal 'm'
WTMP_naive <- stan_naive(train_WTMP, seasonal = TRUE,
                         stepwise = TRUE, m = 52, iter = 1000)
```

```
check_residuals(WTMP_naive)
```

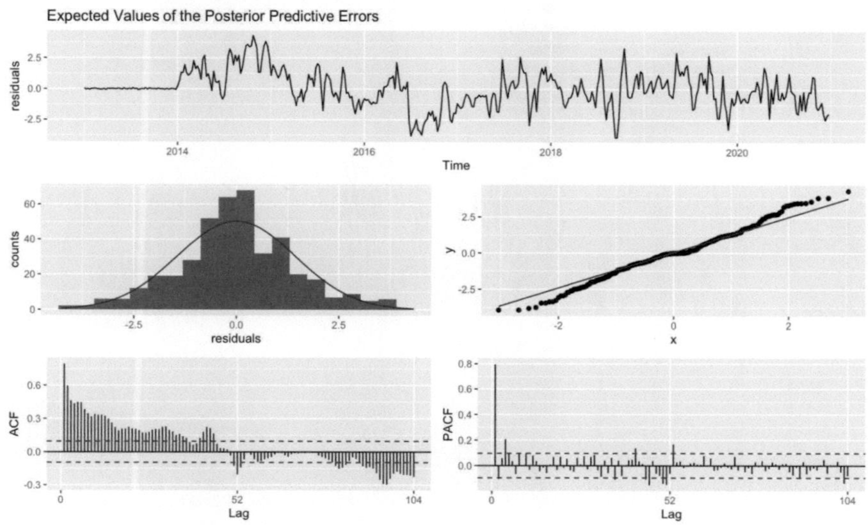

Fig. 7.6 Residual analysis of Random Walk model

Both the ACF and PACF plots (two panels at the bottom of Fig. 7.6) display some level of systematic autocorrelation. A potential improvement to the random walk model would be the inclusion of an AR() term to capture that feature.

7.8 Forecast

The next step, as we have done in the earlier section of the Time Series chapter, is to select the model we will use to forecast the series beyond the observed values. Before we complete that stage, we will forecast values using all the models we built over the time period of the testing set. That way, we will be able to determine which one performs best. As we have done before, we will use graphic and numerical approaches.

```
#Forecast models
WTMP_sarima_for <- forecast(object = WTMP_sarima, h=100)
WTMP_auto_for <- forecast(object = WTMP_auto, h=100)
WTMP_ssm_for <- forecast(object = WTMP_ssm, h=100)
WTMP_naive_for <- forecast(object = WTMP_naive, h=100)
```

```
# ## Full series
plot(train_WTMP, lwd=1.25, lty=1, col="gray4",
    main="", ylab="Water temperature", xlab="Time",
    ylim = c(8, 20), xlim=c(2013,2023), xaxt = "n")
axis(1, at = seq(2013, 2023))
lines(test_WTMP, col="gray4", lwd=1.25, lty=3)
lines(WTMP_naive_for$mean, col="indianred", lwd=1.25)
lines(WTMP_ssm_for$mean, col="green4", lwd=1.25)
lines(WTMP_sarima_for$mean, col="purple3", lwd=1.25)
lines(WTMP_auto_for$mean, col="royalblue", lwd=1.25)
```

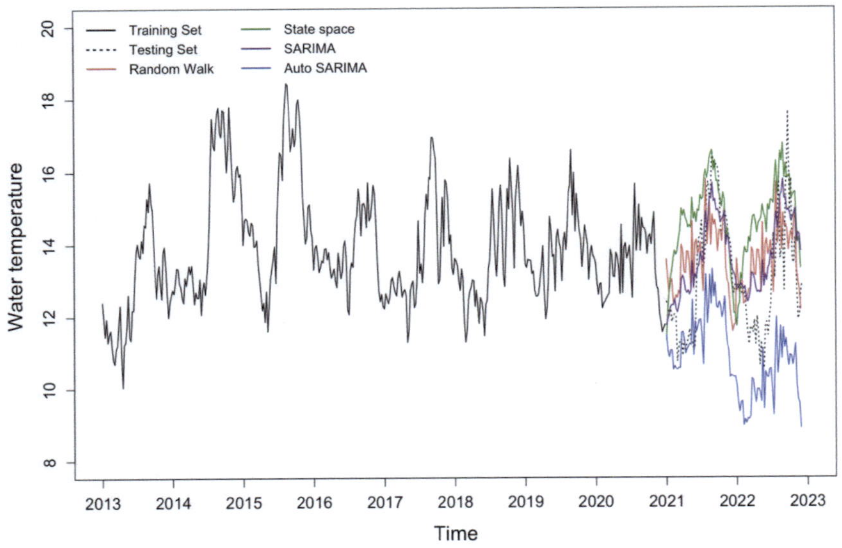

Fig. 7.7 Training set, testing set and forecast from different models

```
legend("topleft", lwd=2, col=c("gray4", "gray4","indianred", "green4", "purple3
    ", "royalblue"),
    lty = c(1,3,1,1,1,1),
    legend=c("Training Set","Testing Set", "Random Walk", "State space", "
        SARIMA","Auto SARIMA" ),bty="n", ncol=2)
```

The plot in Fig. 7.7 shows the entire series (with separate graphic representations for the training and testing sets) and the forecast generated by each model. We can appreciate some of the differences in modeling results, but it is difficult to see which model fits the testing set the best. Now, we zoom in to see the differences between the testing set and each forecast. We can see some of the details of the forecast values but it is clearly not enough to make a final determination as to which model provides the best fit.

We also plot a detailed view in Fig. 7.8 that makes the comparison between the testing set and each model's forecast much easier.

```
## Testing set
plot(test_WTMP, lwd=1.25, lty=3, col="gray4",
    main="", ylab="Water temperature", xlab="Time",
    ylim = c(8, 20), xlim=c(2021,2023), xaxt = "n")
axis(1, at = seq(2021, 2023))
lines(WTMP_naive_for$mean, col="indianred", lwd=1.25)
lines(WTMP_ssm_for$mean, col="green4", lwd=1.25)
lines(WTMP_sarima_for$mean, col="purple3", lwd=1.25)
lines(WTMP_auto_for$mean, col="royalblue", lwd=1.25)
legend("topleft", lwd=2, col=c("gray4","indianred", "green4", "purple3", "
    royalblue"),lty = c(3,1,1,1,1),
    legend=c("Testing Set", "Random Walk", "State space", "SARIMA","Auto
        SARIMA" ),bty="n", ncol=2, cex=1.2)
```

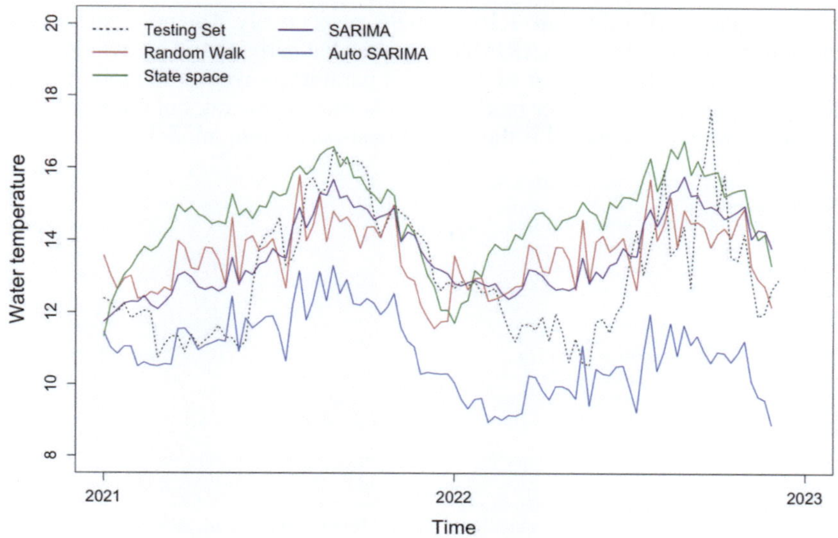

Fig. 7.8 Testing set and forecast from different models

Numeric measures of accuracy

Although charts are useful in supporting our intuition, the final determination of the predictive capacity of each model will come from numeric parameters. In this section we use built-in functions in R packages to generate all the metrics we need for assessing each model's performance. From Table 7.1, we focus our attention on the second (RMSE) and third columns (MAE). These are the *root mean squared errors* and the *mean absolute errors*. According to both measures, the SARIMA(1,0,1)(1,1,1) performs better than any of the other three, confirming that our intuition was correct.

Table 7.1 Accuracy measures of different Bayesian models

	ME	RMSE	MAE	MPE	MAPE	MASE	ACF1	Theil's U
Stan SARIMA method								
Training set	0.04	0.72	0.53	0.09	3.79	0.44	0.01	
Test set	–0.53	1.20	0.98	–4.92	7.81	0.81	0.67	1.65
auto SARIMA method								
Training set	0.00	0.79	0.57	–0.14	4.05	0.47	0.01	
Test set	2.22	2.65	2.28	16.16	16.71	1.90	0.74	3.25
State Space method								
Training set	–0.02	0.69	0.50	-0.29	3.63	0.42	0.06	
Test set	–1.58	2.21	1.82	-13.39	15.02	1.51	0.80	3.18
Random Walk naïve method								
Training set	0.00	1.42	1.05	–0.41	7.47	0.88	0.80	
Test set	–0.46	1.59	1.33	–4.84	10.56	1.11	0.71	2.18

The RMSE and MAE values are 1.16 and 0.96, respectively. The poor performance of the automatic model (auto SARIMA) is somewhat surprising. One would expect that the automatic identification of a model's parameters would be more efficient, and that it would produce better models. That is true in general, but not in all cases. The random walk/naïve model is the second-best-performing model.

```
# Creates the list of model forecast
forecast_objects <- list(
  "Standard SARIMA" = WTMP_sarima_for,
  "Auto SARIMA" = WTMP_auto_for,
  "State Space" = WTMP_ssm_for,
  "Random Walk Naive" = WTMP_naive_for
)

# Calculate all accuracy metrics at once
table.accuracy <- do.call(rbind,
                         lapply(forecast_objects, function(x)
                           forecast::accuracy(x, test_WTMP))
)
```

```
knitr::kable(table.accuracy, digits = 2,
          caption = "Accuracy measures of different Bayesian models") |>
  kableExtra::group_rows(group_label = "Stan SARIMA method", start_row = 1,
       end_row = 2) |>
  kableExtra::group_rows(group_label = "auto SARIMA method", start_row = 3,
       end_row = 4) |>
  kableExtra::group_rows(group_label = "State Space method", start_row = 5,
       end_row = 6) |>
  kableExtra::group_rows(group_label = "Random Walk naïve method", start_row =
       7, end_row = 8) |>
  kable_styling(latex_options = c("HOLD_position","scale_down"))
```

In the end, it seems that the intuition of modeling the series with SARIMA(1,0,1)(1,1,1) paid off, as it is the model that performs best. We will use this model to forecast the values beyond the original dataset.

7.9 Forward Forecast

Finally, after selecting the model with the best fit (based on the smallest RMSE), we forecast values two years into the future (beyond the full dataset). We will use the SARIMA (1, 0, 1) (1, 1, 1) model to estimate values for the next two years of the series. In the code below, we refer back to the complete time series (`WTMP_weekly_TS`) and apply the modeling parameters we used before with the training dataset. The parameter `h=104` specifies the time horizon, that is, 104 periods into the future, which corresponds to two full years.

```
WTMP_sarima_full <- stan_sarima(ts = WTMP_weekly_TS, order = c(1,0,1),
               seasonal= c(1,1,1), prior_mu0 = student(mu=0, sd=1,df=7))
WTMP_sarima_full_for <- forecast(object = WTMP_sarima_full, h=104)
```

We also produce the corresponding plot and show it in Fig. 7.9.

```
plot(WTMP_weekly_TS, lwd=1.25, lty=1, col="gray4",
     main="", ylab="Water temperature", xlab="Time",
    ylim = c(8, 20), xlim=c(2013,2025), xaxt = "n")
axis(1, at = seq(2013, 2025))
```

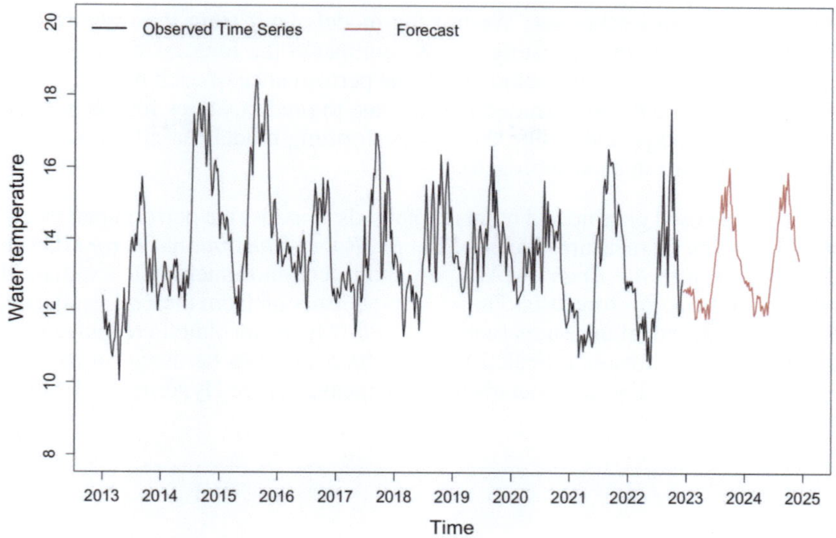

Fig. 7.9 Observed data and forecast SARIMA (1, 0, 1) (1, 1, 1)

```
lines(WTMP_sarima_full_for$mean, col="indianred", lwd=1.25)
legend("topleft", lwd=2, col=c("gray4", "indianred"),
     legend=c("Observed Time Series","Forecast"),bty="n", ncol=2)
```

7.10 Workflows in Time Series Analysis

1. **Data preparation**: We read the data files, cleaned them, and created a time series object that is suitable for further analysis. One fundamental aspect to consider is that time series techniques require uniform intervals for all observations. For *extrapolation* models, we used the sea level dataset (although the data were not converted into time series (*ts*) objects. In the case of *ARIMA* models, we used monthly data on housing permits. For *Bayesian* models, we used weekly data on ocean water temperature.
2. **Descriptives**: For data with more complex behaviors, the most common practice to understand a time series is to *decompose* it into its trend, seasonality, and residuals components.
3. **Training and testing sets**: We split the dataset into training and testing sets to build models and evaluate their performance.
4. **Build Appropriate Models**: In each of the sections, we built different models to capture the behavior of the data and create meaningful forecasts. In the case of the ARIMA models, we built ARIMA(), autoARIMA, and Holt-Winter's models. Then we used internal functions of the `forecast` package to build the *mean*, *naive*, *drift*, and *seasonal naïve* models. In the case of Bayesian models, we used the `bayesforecast` package to build several models, including *Stan ARIMA*, *Auto SARIMA*, *State Space*, and *Random Walk (Naïve)* models.

5. **Forecasting the testing set**: We use the models built from the training set to generate forecasts of the testing set. We compared the forecast values with the actual values in the testing set to assess the performance of each model.
6. **Forward forecast**: We extended the forecast to predict values for the next corresponding time periods using the best-performing model (based on the lowest measure of error in each approach).

In all cases, we used graphic and numeric methods to assess the performance of each model. The numeric measures included the R^2, *Root Mean Squared Error* (RMSE), and the *Mean Absolute Error* (MAE), which are common metrics to evaluate the accuracy of time series forecasts. The `accuracy` function from the `forecast` package provides additional measures such as MAPE (Mean Absolute Percentage Error) and MASE (Mean Absolute Scaled Error), which can also be useful in assessing model performance. For more details on these measures, see Hyndman and Athanasopoulos [7].

References

1. Cox, David R. 2006. *Principles of Statistical Inference*. Cambridge, UK: Cambridge University Press. https://doi.org/10.1017/CBO9780511615869.
2. Thomas, J. 2025. *Faulkenberry*. Bayesian Statistics: The Basics. Routledge. https://doi.org/10.4324/9781003470199.
3. Gelman, Andrew, et al. 2013. *Bayesian Data Analysis*, 3rd edn. Chapman and Hall/CRC. https://doi.org/10.1201/b16018.
4. McElreath, R. 2020. *Statistical Rethinking: A Bayesian Course with Examples in R and STAN*, 2nd edn. Chapman and Hall/CRC. https://doi.org/10.1201/9780429029608.
5. National Data Buoy Center. 2025. *NDBC Station 46042 Monterey, CA*. https://www.ndbc.noaa.gov/station_page.php?station=46042. Accessed Oct 2024.
6. Scott, Steven, and Hal Varian. 2014. Predicting the present with Bayesian structural time series. *International Journal of Mathematical Modelling and Numerical Optimisation* 5: 4–23. https://doi.org/10.1504/IJMMNO.2014.059942.
7. Hyndman, Rob J., and George Athanasopoulos. 2018. *Forecasting: Principles and Practice*. English, 2nd edn. Australia: OTexts.
8. Peng, Roger. 2020. *A Very Short Course on Time Series Analysis*. https://bookdown.org/rdpeng/timeseriesbook/.
9. Abadi, Martín, et al. 2016. *TensorFlow: Large-Scale Machine Learning on Heterogeneous Systems*. Software available from tensorflow.org. https://arxiv.org/abs/1603.04467.
10. Brodersen, Kay H., et al. 2015. Inferring causal impact using Bayesian structural time-series models. *The Annals of Applied Statistics* 9(1): 247–274. https://doi.org/10.1214/14-AOAS788.

Part IV
The Effect of Location

Chapter 8
Spatial Analysis

8.1 A Brief Introduction

All human activity occurs in space—whether on land, water, or even in extraterrestrial environments. These activities, and the spaces in which they occur, are deeply interrelated. Understanding the spatial dimension of human actions–why they occur in specific places, how they change over time, and what implications they carry–is the central objective of spatial analysis. This field intersects with numerous scientific disciplines and areas of professional practice, including geography, demography, spatial econometrics, transportation planning, and land-use management. Each of these areas both influences and is influenced by spatial understanding, particularly in relation to human-environment interactions.

Recent decades have witnessed a dramatic increase in the availability of spatially referenced data (also called geo-referenced data). The same algorithms that allow people to find friends via smartphones, navigate unfamiliar cities, or track deliveries are also generating vast amounts of spatial data. Much of this data is stored by internet service providers (ISPs), logistics operators, and telecommunication companies, and, increasingly, made available to researchers and policy analysts. While such developments raise valid privacy concerns, they also present significant opportunities for understanding and addressing spatial dimensions of climate adaptation.

From foundational theorists like Moran [1], Geary [2], Tobler [3], and Anselin [4] to more recent advances leveraging machine learning [5], spatial analysis has evolved to encompass a wide array of analytical approaches and tools. This chapter presents a focused selection of these tools that are most relevant to climate change adaptation analytics.

A critical function of spatial information is its ability to integrate otherwise disjointed datasets. For example, weather forecasts are often translated into estimates of population at risk by connecting the geospatial coordinates of forecasted events with demographic data (e.g., county-level population counts in the United States). Such applications demonstrate the utility of spatial analysis in both communication and decision-making.

Spatial data analysis becomes even more powerful when temporal dynamics are incorporated. Climate adaptation efforts frequently rely on tracking changes over time in the same geographic units, allowing for trend analysis and scenario model-

F. DePaolis, *Workflows for Climate Change Adaptation Analytics*,
https://doi.org/10.1007/978-3-032-14004-3_8

ing. These spatiotemporal datasets provide the empirical foundation for local policy design and evaluation. Temporal analysis is addressed in further detail in the chapter on Time Series Analysis.

In this chapter, we begin by introducing the dataset used throughout our spatial analyses and demonstrating how to construct various types of maps. We then examine spatial autocorrelation and spatial dependency using common statistical tools. Finally, we briefly explore how to use R to export shapefiles for integration with other Geographic Information Systems (GIS) and other extensions of spatial analysis.

8.2 Measuring Spatial Relationships

Proximity and Its Definitions

Waldo Tobler's first law of geography states: "Everything is related to everything else, but near things are more related than distant things" [3]. While intuitive, this principle demands precise operationalization in empirical work. Proximity may refer to physical closeness, but in many applications, it must also consider economic, social, or even political connections.

Consider physics: heat transfer via conduction requires direct contact between molecules, whereas quantum entanglement allows influence across vast distances. Similarly, in international trade, geographic closeness does not always predict the intensity of economic exchange—South Korea trades far more with the United States than with its neighbor, North Korea. Hence, physical proximity, while often necessary, is not always sufficient to establish relational influence.

In network analysis, proximity is often expressed through an adjacency matrix. In the simplest form, such a matrix contains binary values indicating the presence or absence of a connection. More complex forms use weighted, directed graphs to capture the strength and direction of relationships.

Conceptualizing Spatial Processes

In spatial analysis, proximity must support the modeling of interactions across units and their attributes. Haining [6] organizes these interactions into four primary processes: diffusion, exchange, interaction, and dispersal.

- *Diffusion* refers to the spread of an attribute (e.g., air quality) across space. For example, particulate matter from industrial pollution may drift and affect regions far from the source.
- *Exchange* denotes economic or informational flows, such as trade between urban centers or the transfer of technology.
- *Interaction* describes mutual influence between regions, as with the coastal effects of seawall construction on neighboring beaches.
- *Dispersal* involves the physical movement of people, animals, or plants across space.

These processes are highly relevant in climate adaptation, where impacts and responses often propagate across spatial boundaries. This clearly reveals the direct implications for technical, regulatory, and political actions at all government levels.

Spatial Weights Matrices

The primary instrument for encoding spatial relationships is the Spatial Weights Matrix (SWM). This is a square matrix with non-negative values representing the degree of spatial association between units. Diagonal entries are typically zero, reflecting that a spatial unit is not a neighbor to itself.

Numerous methods exist for constructing SWMs, including contiguity-based approaches (e.g., rook or queen adjacency), distance thresholds, and k-nearest neighbors. For further technical details, see Getis and Ord [7], Getis and Aldstadt [8], and Fischer and Getis [9].

Distance-based methods define spatial relationships based on the geographic distance between observations. A **distance threshold** approach designates two spatial units as neighbors if the distance between their centroids (or other defined points within the units) falls below a pre-specified cutoff distance. This method is beneficial when spatial interactions are likely to diminish rapidly with increasing distance. Determining an appropriate distance threshold is critical and often involves balancing the desire to capture local interactions with the need to ensure sufficient connectivity across the study area.

The **k-nearest neighbors** (KNN) method defines spatial relationships by identifying a fixed number of nearest neighbors for each spatial unit, irrespective of the actual distance. For a given spatial unit, its k closest neighbors are identified based on the shortest distances to other units. This approach ensures that every spatial unit has the same number of neighbors, which can be advantageous in specific analytical contexts, especially when dealing with irregularly spaced data or areas with varying densities of observations.

In R, the `spdep` package [10] provides versatile tools for creating and analyzing spatial neighbors, especially using polygon data. It supports various neighborhood definitions, from simple adjacency to complex multi-order relationships. The `SDPDmod` package [11] offers further extensions for dynamic spatial modeling.

Classical Spatial Statistics

Several key measures have been foundational in the development of spatial statistics:

- **Moran's I**: A measure of global spatial autocorrelation developed by Moran [12], later generalized by Moran [1], and popularized by Cliff and Ord [13].
- **Geary's C**: Also known as the contiguity ratio, developed by Geary [2], which emphasizes local spatial variation.
- **Getis-Ord G**: Introduced by Getis and Ord [7], this measure identifies local clusters of high or low values.

For practical applications and advanced modeling, see Anselin [4], who introduced the Local Indicators of Spatial Association (LISA), enabling localized spatial diagnostics.

Machine Learning in Spatial Analysis

A novel and promising development is the use of machine learning for spatial inference. Agarwal et al. [5], working with researchers at Google, introduced the Population Dynamics Foundation Model (PDFM). This model performs geospatial interpolation to predict socio-economic, health, and environmental indicators in data-sparse regions. The increasing availability of high-resolution spatial data, combined with powerful machine learning algorithms, signals a transformative shift in the field of spatial analytics. This is now implemented in Python, but it is very likely that there will be further developments (github.com/google-research/population-dynamics).

8.3 Distribution Over Space: The Case of Air Pollution

Air pollution is one of the most pervasive environmental stressors with direct implications for public health and climate adaptation strategies. A particularly harmful form of air pollution is airborne particulate matter (PM), which is a heterogeneous mixture of solid and liquid particles suspended in the atmosphere. According to the California Air Resources Board (CARB), PM varies considerably in size, shape, and chemical composition, encompassing a broad range of substances including inorganic ions, elemental carbon, metallic compounds, organic compounds, and crustal materials [14].

CARB and other regulatory agencies classify PM based on aerodynamic diameter. Particles with a diameter of 10 μm or less (PM_{10}) are considered inhalable and can penetrate the upper regions of the respiratory tract. A more concerning subset, fine particulate matter ($PM_{2.5}$), is particles with a diameter of 2.5 μm or smaller. Due to their size, $PM_{2.5}$ can infiltrate deep into the lungs, reaching the alveolar sacs, and is associated with a wide array of health problems, including respiratory and cardiovascular diseases, and increased mortality (Pope and Dockery [15] and World Health Organization [16]).

Particulate matter originates from both natural and anthropogenic sources and can be classified as either primary or secondary. Primary PM is emitted directly into the atmosphere through processes such as industrial activities, vehicular emissions, construction, and biomass burning (e.g., wildfires). Secondary PM forms through complex atmospheric chemical reactions involving precursor gases like sulfur dioxide (SO_2), nitrogen oxides (NO_x), and volatile organic compounds (VOCs). These secondary reactions contribute significantly to urban haze and overall $PM_{2.5}$ concentrations [17].

Climate and Air Pollution Interactions

The interactions between climate change and air pollution are increasingly recognized in the scientific literature. Elevated temperatures, intensified heatwaves, reduced precipitation, and increased desertification are all contributing to the frequency and severity of air pollution events. Wildfires—exacerbated by climate change—emit vast quantities of primary particulates and precursor gases, significantly deteriorating air quality and increasing respiratory health risks (Reid et al. [18] and World Meteorological Organization [19]).

In light of these interactions, air pollution control is not only a mitigation strategy but also an adaptation measure within the climate policy framework [20]. Reducing PM emissions can decrease climate-related vulnerabilities and improve public health, particularly in densely populated or disadvantaged areas where exposure is disproportionately high. Therefore, spatial analysis of air pollution is essential for identifying hotspots, understanding exposure patterns, and informing climate-resilient urban planning and public health interventions. This section applies spatial analytics methods to explore the geographic distribution of air pollution as a foundation for modeling and decision-making.

8.4 Air Pollution Data

For this analysis, we use data from the U.S. Centers for Disease Control and Prevention (CDC). The dataset titled Daily Census Tract-Level PM2.5 Concentrations, 2016–2020, contains nearly 140 million observations and offers high spatial and temporal resolution suitable for environmental health and spatial analytics.

To demonstrate a specific use case, we extract a subset of this dataset for the state of California, using observations from July 15, 2019. The relevant fields used for sub-setting were `date` and `statefips` (the latter denoting state-level identifiers based on the Federal Information Processing Series−FIPS−codes). The resulting subset contains 8,040 records at the census tract level.[1]

This dataset, available through the companion website to this book, can be directly downloaded and queried to suit various analytical objectives. We encourage readers to replicate and adapt the extraction process to gain familiarity with handling large-scale environmental datasets. For an alternative approach applied to Western states in the United States, see Reid et al. [21].

The CDC notes that the dataset provides "estimates of the mean predicted concentration and associated standard error" of $PM_{2.5}$ levels, derived through a data fusion process that combines satellite-derived aerosol optical depth, chemical transport models, and ground-based monitoring data [22]. For our analyses, we use the $PM_{2.5}$ concentration estimates as the principal variable of interest. Since the data was acquired in .csv format, it requires no additional preprocessing.

```
PM2_5_Jul15 <- read_csv("./data/PM2.5_Concentrations__15JUL2019_Calif.csv")
```

8.5 Geographic Data

In this chapter, we work with georeferenced data files—data objects that contain an explicit geographic dimension, along with the inherent spatial relationships among geographic units, also known as topology. Many readers may be familiar with Geographic Information Systems (GIS) and the spatial data structures commonly used, such as shapefiles.

The analysis presented here operates at both the county and census tract levels. Consequently, we must associate our variable of interest—fine particulate matter ($PM_{2.5}$)—with specific geographic locations. The $PM_{2.5}$ data include latitude and longitude coordinates that correspond to the locations for which the $PM_{2.5}$ concentrations were estimated. We align these coordinates with the spatial units represented in the county and census tract shapefiles.

[1] There are 8,057 census tracts in the state of California.

To download the corresponding geographies for the State of California, we utilize the `counties` and `tracts` functions from the R package `tigris` [23]. In the case of tracts, we use `cb=FALSE` to obtain the most detailed TIGER line map, and filter out units with land area equal to zero.

Counties

```
CA_counties <- counties("CA", cb = FALSE) %>%
  mutate(centroids = st_centroid(st_geometry(.)))
```

Tracts

```
# Gets files from TIGRIS and Removes areas with ALAND=0 -> polygons in the
      water with zero land area
CA_tracts <- tracts("CA", cb = FALSE) %>%
  mutate(centroids = st_centroid(st_geometry(.))) %>%
  filter(ALAND!='0')
```

8.6 Mapping

Despite the rise in complex statistical modeling and big data analytics, maps remain a powerful and intuitive tool for spatial data visualization. In parallel with the expansion of spatial data availability, R packages have evolved to allow for the processing of both vector and raster data, as well as to create sophisticated maps of superb cartographic quality. As a special category of visualization (see Chap. 3), maps enable an immediate and intuitive understanding of spatial variation and distribution patterns. With the extensive array of mapping packages available in R, we are afforded flexibility in how we present georeferenced data. The challenge lies not in access to tools, but in making deliberate choices that serve the clarity and purpose of the analysis.

This section demonstrates several approaches to mapping $PM_{2.5}$ concentration data across different spatial resolutions.

A note on geographic data types

Vector data represents geographic features using points, lines, and polygons defined by mathematical equations (x, y coordinates), providing precise spatial representations for discrete features with well-defined boundaries like roads, rivers, and property lines. **Raster** data represents geographic information as a grid of equally sized cells or pixels, each containing a value that represents a particular attribute or phenomenon like elevation, temperature, or land use (gisgeography.com).

8.6.1 Initial Visualization: Point Map

Our first visualization involves plotting $PM_{2.5}$ data points using their geographic coordinates. The map in Fig. 8.1 is a scatter of dots that resembles the map of California. Though rudimentary, this visualization already reveals discernible spatial patterns, including the rough outline of the state and denser clusters of pollution in urbanized regions and valleys.

Fig. 8.1 Pollution points using coordinates from dataset

While the map lacks overlays such as political boundaries or topographic context, it provides a preliminary sense of spatial distribution. If unfamiliar with California's geography, readers may still infer major human settlement patterns or the presence of natural features such as the coastal ranges or the Sierra Nevada, which appear as vast empty areas in the map.

The coordinate reference system (CRS) used in this visualization is WGS 1984 (EPSG 4326), which specifies the data in terms of geographic coordinates (latitude and longitude). The function `st_as_sf` from the `sf` package converts a data frame object into an `sf` object, which contains a geometry column holding spatial data. This is a critical step in creating a georeferenced dataset.

```
pollut_points = st_as_sf(PM2_5_Jul15, coords = c("latitude", "longitude"), crs
     = 4269)

plot(st_geometry(pollut_points), pch=16, col="steelblue", cex=0.2)
```

8.6.2 *Adding County Boundaries*

To enhance spatial contextualization, we overlay county boundaries onto the existing point map and create the map in Fig. 8.2. This allows for a clearer understanding of how data collection points and estimated pollution levels are distributed across political jurisdictions.

```
map <- ggplot() +
  geom_sf(data = CA_counties, fill = NA, color = "black") +
  geom_sf(data = pollut_points, size = 0.25, shape = 16, color = "steelblue") +
  theme_minimal(base_size = 8) +
```

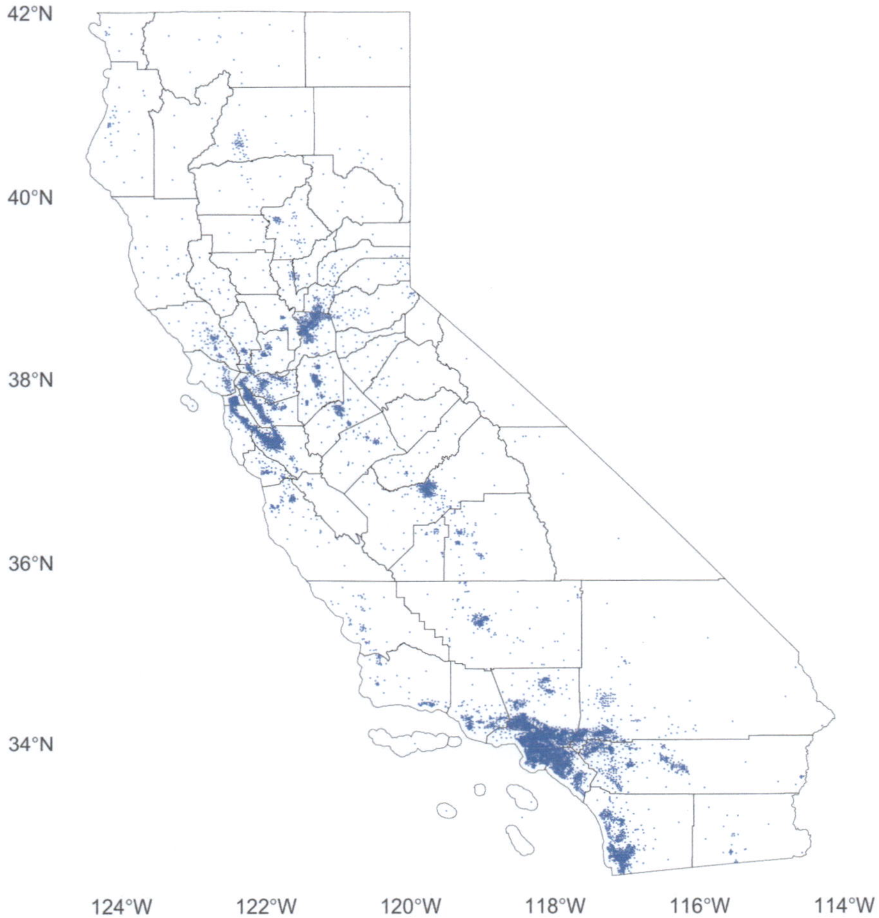

Fig. 8.2 Pollution points with county boundaries

```
  theme(
    axis.title = element_blank(),
    axis.text = element_text(size = 5),
    panel.grid = element_line(color = "white", linewidth = 0.25)
  )

map
```

We can further enrich the map by classifying the $PM_{2.5}$ points according to their estimated concentrations. In our dataset, these values range approximately 1.7–17.5 μ g/m^3 (micrograms per cubic meter). This classification helps distinguish areas of low, moderate, and high pollution intensity, as shown in Fig. 8.3.

We can also restrict the view to a specific region—such as Southern California—where $PM_{2.5}$ levels appear to be more elevated, thereby offering a more focused analysis. We have defined the Southern California region between 116.5° and 119° West, and 32° and 34.5° North.

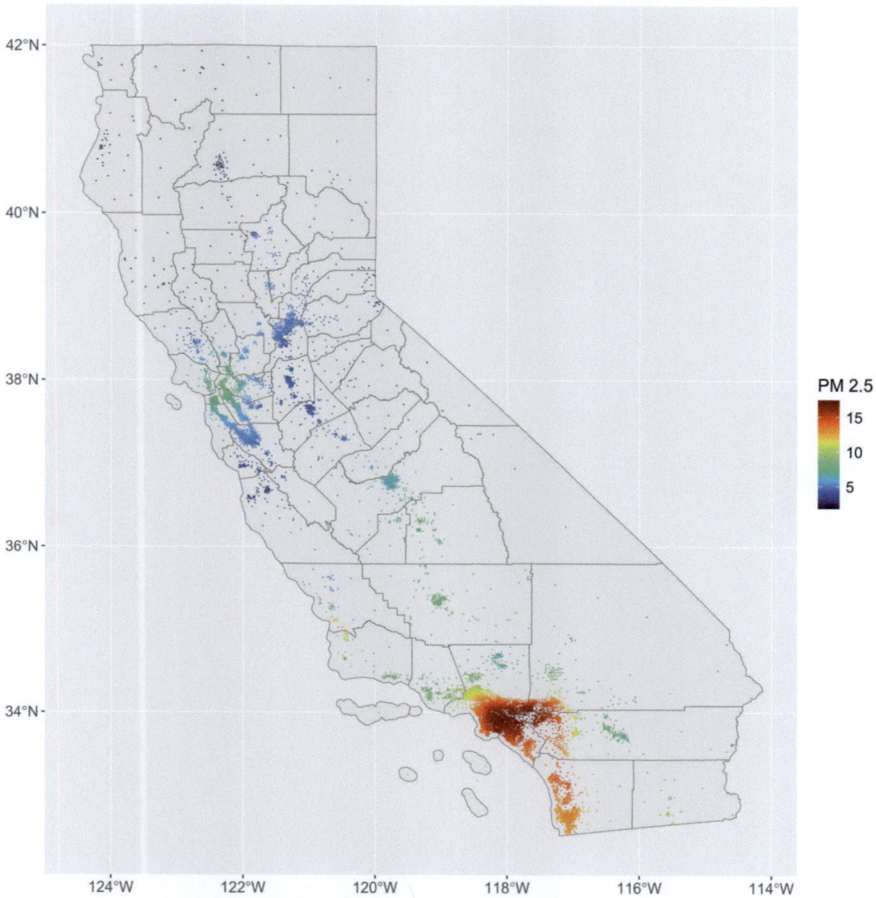

Fig. 8.3 Pollution points colored by PM2.5 level with county boundaries

```
pp1 <- ggplot(data = CA_counties) +
  geom_sf() +
  geom_sf(data = pollut_points,
        aes(color = DS_PM_pred),
        size = 0.25,
        shape = 16) +
  scale_color_viridis_c(option = "turbo", name = "PM 2.5") +
  theme(
   axis.title = element_blank(),
   axis.text = element_text(size = 5),
   panel.grid = element_line(color = "white", linewidth = 0.25),
   legend.text = element_text(size = 6),
   legend.title = element_text(size = 7)
  )
pp1
```

In Fig. 8.4, we show the Southern California area with pollution points colored by $PM_{2.5}$ levels.

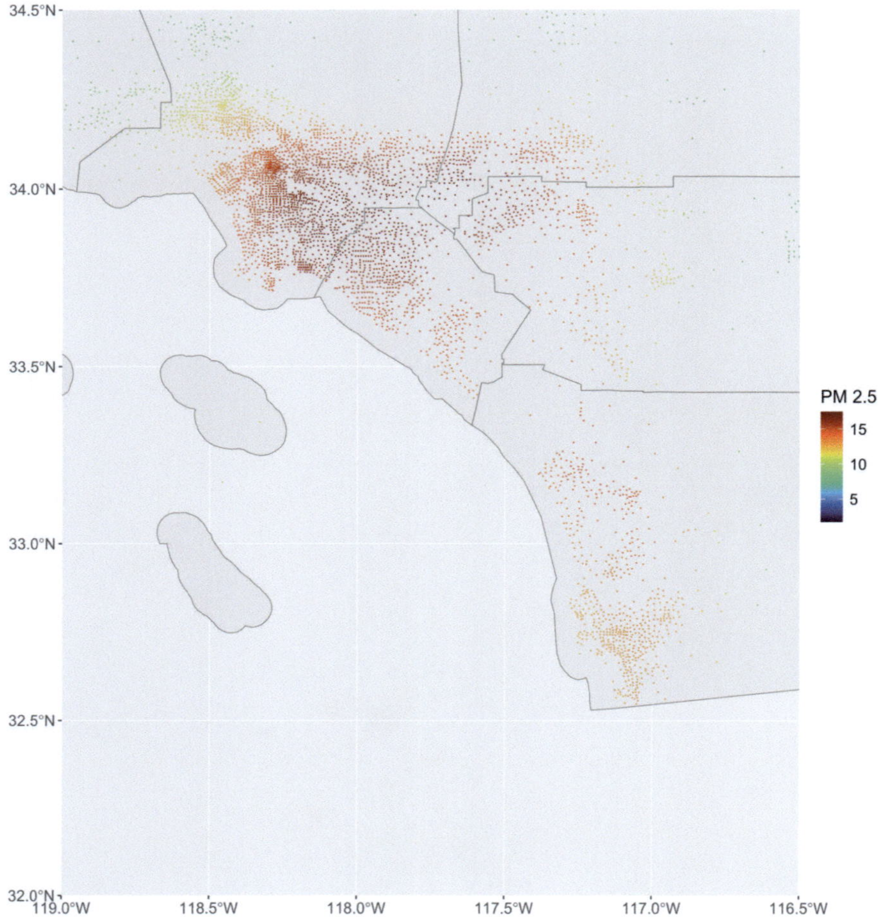

Fig. 8.4 Pollution points colored by PM2.5, enlarged for a subregion

```
pp1 + coord_sf(xlim = c(-119,-116.5), ylim=c(32,34.5), expand=FALSE)
```

8.6.3 Mapping at the Census Tract Level

To represent pollution exposure across entire geographic units, we aggregate the data to the census tract level. This process involves merging the georeferenced $PM_{2.5}$ point data with the census tract shapefiles using the `st_join` function from the `sf` package [24]. Once merged, we can create choropleth maps where each tract is colored according to the $PM_{2.5}$ concentration recorded within its boundaries. If there are multiple data points for each spatial unit, it is possible to select the type of computation or value selection, such as maximum, average, sum, etc.

This approach provides a clearer picture of spatial heterogeneity and is particularly useful for identifying environmental justice concerns, as socio-economically vulnerable populations often reside in areas with higher pollution levels [25].

Notice that Modoc County, in the top right-most corner of the California map has four census tracts. Two adjacent points in the center of Modoc County may suggest that there are more than one point in the census tract, when in reality there is a census tract with a very small area around the town of Alturas, the largest town in the county. There are similar cases where the dot showing the data collection location obscures the boundaries of census tracts of small extent.

We see this map in Fig. 8.5.

```
pp2 <- ggplot(data = CA_tracts) +
  geom_sf() +
  geom_sf(data = pollut_points,
          aes(color = DS_PM_pred),
          size = 0.5,
          shape = 16) +
  scale_color_viridis_c(option = "turbo", name = "PM 2.5") +
  theme(
    axis.title = element_blank(),
    axis.text = element_text(size = 5),
    panel.grid = element_line(color = "white", linewidth = 0.25),
    legend.text = element_text(size = 6),
    legend.title = element_text(size = 7)
  )

pp2
```

The second map (Fig. 8.6) shows the same information but for a subset corresponding to Southern California.

```
pp2 + coord_sf(xlim = c(-119,-116.5), ylim=c(32,34.5), expand=FALSE)
```

Finally, we can generate maps in which entire geographic units—such as census tracts, as used in the examples below—are shaded or colored based on the air quality measurements recorded at specific point locations within their boundaries. This approach offers a synthetic representation of spatial variation, as we typically have only a single data point per tract, or in some cases, none at all. Consequently, the assumption that a single measurement reflects air quality uniformly across the entire tract can introduce inaccuracies. This limitation may be negligible in small, densely populated tracts where pollution levels are more spatially homogeneous. However, in large, sparsely populated tracts, this simplification can lead to significant distortions. For instance, if a particulate matter reading is taken near a localized pollution source—such as a highway, industrial facility, or mining operation—using that value to characterize the entire tract may result in an overestimation of pollution levels in areas farther from the source. Therefore, while tract-level visualizations can be useful for broad spatial assessments, they should be interpreted with caution, especially in heterogeneous or geographically expansive regions.

Figure 8.7 shows the map for the State of California, and Fig. 8.8 maps it for the Southern California region.

```
tracts_points <- st_join(CA_tracts, pollut_points)
```

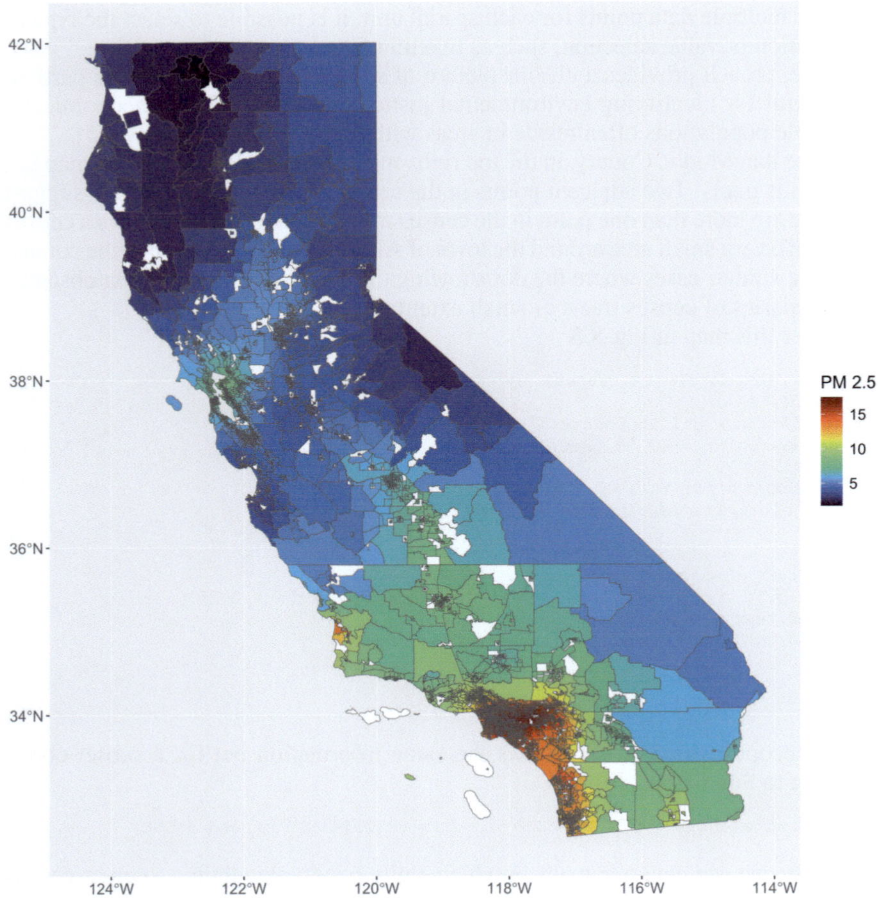

Fig. 8.5 PM2.5 pollution levels with tract boundaries

State

```
pp3 <- ggplot(data = tracts_points) +
  geom_sf(aes(fill = DS_PM_pred)) +
  scale_fill_viridis_c(option = "turbo", name = "PM 2.5", na.value = "white") +
  theme(
    axis.title = element_blank(),
    axis.text = element_text(size = 5),
    panel.grid = element_line(color = "white", linewidth = 0.15),
    legend.text = element_text(size = 6),
    legend.title = element_text(size = 7)
  )
pp3
```

Southern California

```
pp3 + coord_sf(xlim = c(-119,-116.5), ylim=c(32,34.5), expand=FALSE)
```

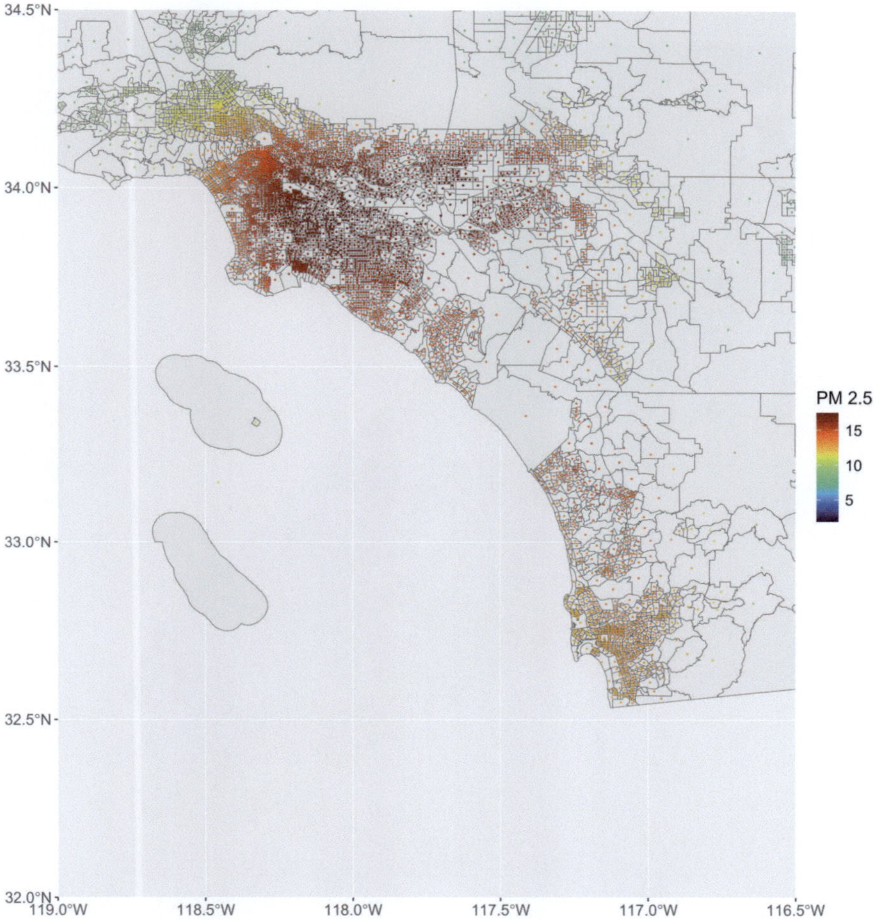

Fig. 8.6 PM2.5 pollution levels with tract boundaries, for Southern California

8.7 Spatial Autocorrelation

In the *Hedonics* section of the *Regression* chapter, we introduced classical measures of association such as Pearson's correlation coefficient, which describes the linear relationship between two continuous variables. While this form of association considers the relationship between *two* distinct variables, correlation can also be considered along a *single* variable, either across time or across space.

Temporal autocorrelation—the correlation of a variable with itself over successive time intervals—was discussed in the *Time Series* chapter. There, we explored the concept of autocorrelation in more depth, focusing on how observations of a single variable at different time points can be statistically dependent, often necessitating models that explicitly account for temporal structure.

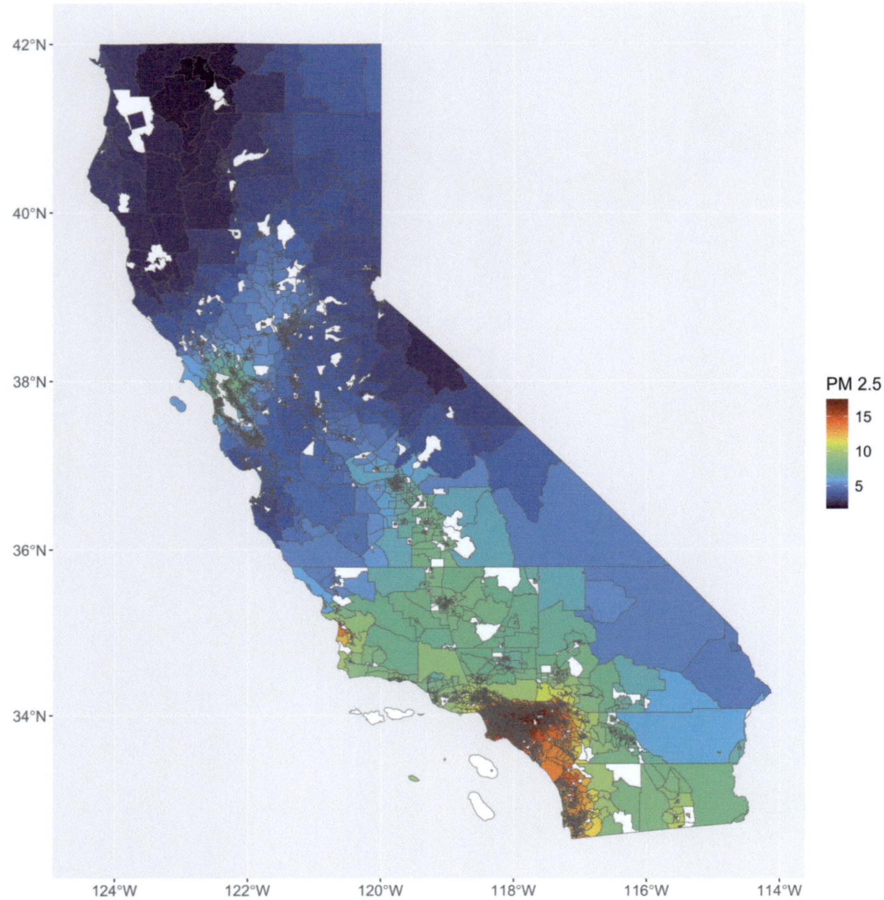

Fig. 8.7 PM2.5 pollution levels at the census tract level

Similarly, **spatial autocorrelation** refers to the degree of dependency among observations of the same variable across spatial units. Unlike temporal autocorrelation, which is inherently ordered by time, spatial autocorrelation depends on how "neighboring" units are defined and how spatial relationships are encoded. Positive spatial autocorrelation indicates that similar values occur near each other more often than expected under spatial randomness, whereas negative spatial autocorrelation suggests a tendency for dissimilar values to be located nearby.

To quantify spatial autocorrelation, a number of statistical measures have been developed, each capturing different aspects of spatial dependency. In this section, we focus on three widely used statistics: Moran's I [12], Geary's C [2], and the Getis-Ord G statistic [7]. Each of these measures requires a specification of spatial structure—typically in the form of a **spatial weights matrix**—to determine which spatial units are considered neighbors and how influence is exerted across space.

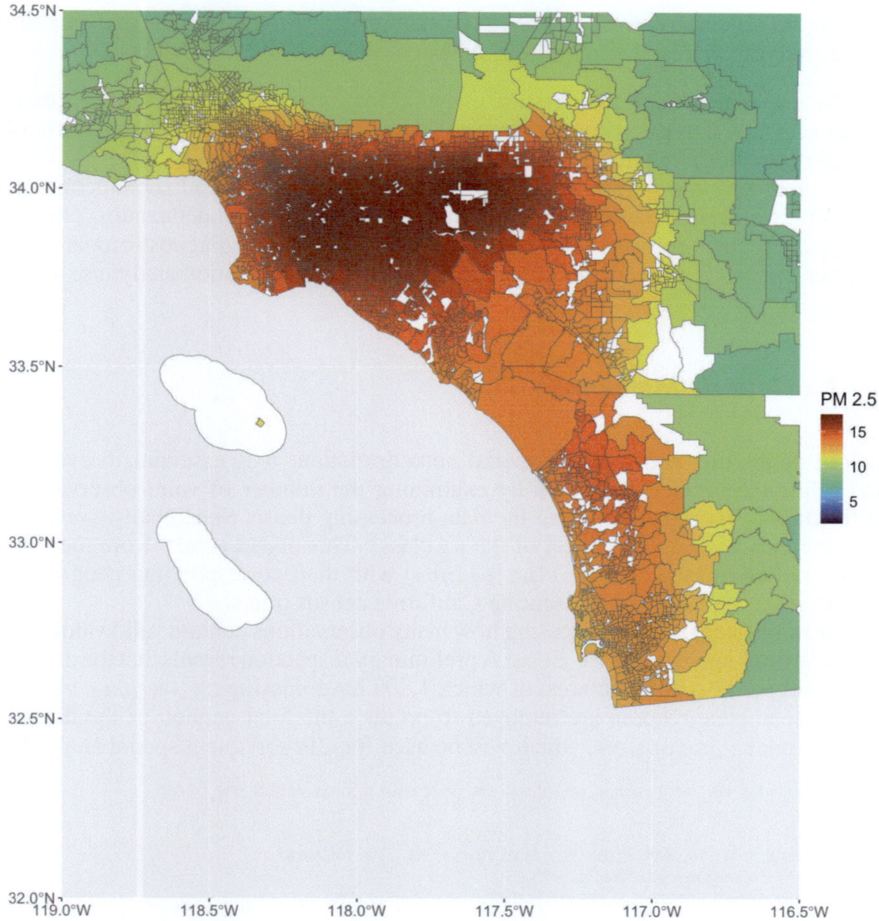

Fig. 8.8 PM2.5 pollution levels at the census tract level, for Southern California

The construction of a spatial weights matrix begins with the definition of neighborhood relationships among spatial units. These relationships can be based on contiguity (shared borders or vertices), distance thresholds, or k-nearest neighbors, depending on the spatial context and research question. A **contiguity matrix** is a representation of such relationships, identifying which spatial units are considered adjacent to each other. For a detailed discussion on spatial weights and neighborhood structures, see Bivand, Pebesma, and Gómez-Rubio [26], Anselin [27], and LeSage and Pace [28].

In R, spatial weights are typically managed using objects of class `listw`, provided by the `spdep` package [29]. A `listw` object is a compound data structure composed of two key components:

- **Neighbors object** (nb): a list defining the neighborhood structure, i.e., which spatial units are adjacent to which others.

- **Weights object**: a corresponding list or matrix of numerical weights that quantify the strength of spatial relationships between each unit and its neighbors.

These spatial weights matrices are essential for computing the spatial autocorrelation statistics introduced in this section, as they formally encode the structure of spatial dependence required by Moran's I, Geary's C, and Getis-Ord G.

By analyzing the spatial autocorrelation of environmental and socio-economic variables, we can detect spatial patterns that may be critical for understanding climate vulnerability and informing adaptation strategies. These spatial diagnostic tools could very well serve as a foundational component of climate adaptation analytics.

8.7.1 Data Refinements

Before computing measures of spatial autocorrelation, it is essential to evaluate and refine the dataset. We begin by examining the number of valid observations in the dataset constructed during the data processing phase. Specifically, we work with the `tracts_points` object of the `sf` class.[2] This object results from merging the air pollution estimates (`pollut_points`) with their corresponding geographic references (`CA_tracts`), representing California census tracts.

The initial step involves assessing how many observations contain valid values for the variable of interest, `DS_PM_pred`. A preliminary inspection reveals that the dataset comprises 9,242 spatial features, of which 1,209 have missing `DS_PM_pred` values. These incomplete records are removed to produce the final version of the dataset named `tracts_points_NA`, which will be used for all subsequent spatial analyses.

```
tracts_points_NA <- tracts_points[!is.na(tracts_points$DS_PM_pred),]
```

```
hist(tracts_points_NA$DS_PM_pred,xlab = "PM 2.5 values",
    ylab = "Frequency", main ="",
    breaks=seq(0,20,1), cex.axis=0.7, cex.lab=0.7)
```

The variable `DS_PM_pred` measures estimated $PM_{2.5}$ concentration levels. The histogram of this variable in Fig. 8.9 reveals a bimodal distribution, with clusters around 5 and 17 ppm. Following our recommendation of best practices in the *Descriptives* chapter, we obtain the summary statistics for the variable of interest.

```
summary(tracts_points_NA$DS_PM_pred)
```

```
## Min. 1st Qu. Median Mean 3rd Qu. Max.
## 1.744 5.901 10.344 10.297 14.645 17.453
```

Next, we define the spatial relationships among geographic units. This requires building a spatial neighbor structure—specifically, a list of polygons and their neighboring polygons. The function `poly2nb` from the `spdep` package generates this neighbor list, where the queen parameter governs adjacency criteria. If `queen = TRUE`, polygons sharing even a single point are considered neighbors (*Queen* contiguity); if `FALSE`, a shared edge is required (*Rook* contiguity requires multiple points).

[2] See: https://r-spatial.github.io/sf/articles/sf1.html.

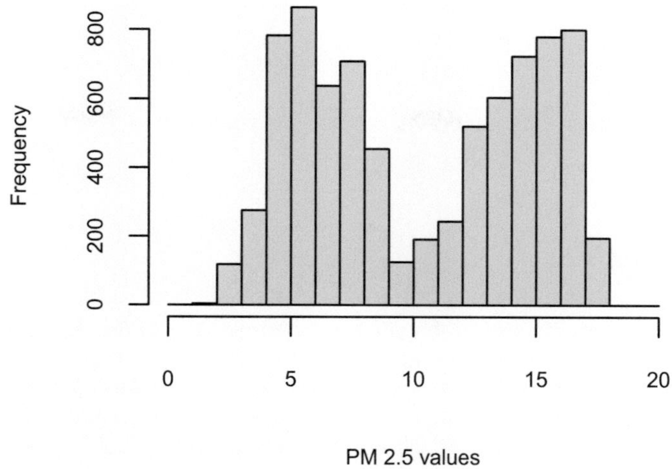

Fig. 8.9 Distribution of PM2.5 Estimated Values–Histogram

Polygons with no neighbors are then identified and removed. This is achieved by the `which()` function (to filter for zero-neighbor polygons). Ten such polygons are identified, likely representing offshore islands along the southern and central California coastline.

An additional data structure is required: the **spatial weights matrix**. This matrix encodes the strength of spatial relationships between neighboring units. Using the `nb2listw` function from the `spdep` package, we generate the matrix with `style =" W"` (row-standardized weights). Although we removed all zero-neighbor polygons, we retain the `zero.policy = TRUE` argument as a reminder for cases where such polygons are retained.

These spatial structures—`pollut_nb` (neighbor list), `pollut_lw` (weights matrix), and `pollut_lag` (spatially lagged values)—form the basis for computing spatial autocorrelation statistics such as Moran's I, Geary's C, and Getis-Ord G. See Pebesma and Bivand [10] for computational and methodological details.

```
#First build of neighbors list
nb <- poly2nb(tracts_points_NA, queen=FALSE) # require more than one shared
    point

# Calculate number of polygons' neighbors and identify them
empty_nb <- which(card(nb) == 0)

# List of polygons with zero neighbors
empty_nb

# Remove polygons with empty neighbor sets from the data
pollut_subset <- tracts_points_NA[-empty_nb, ]

# Identify neighbors with rook contiguity (edge touching)
pollut_nb <- poly2nb(pollut_subset, queen = FALSE)

# Binary weighting assigns a weight of 1 to all neighboring features
# and a weight of 0 to all other features
pollut_lw <- nb2listw(pollut_nb, style="W", zero.policy = TRUE)
```

```
# Calculate spatial lag of PM2.5
pollut_lag <- lag.listw(pollut_lw, pollut_subset$DS_PM_pred)
```

```
## [1] 257 1111 1453 1470 2284 2465 3082 3085 6521 6718
```

8.7.2 Moran's I

Moran's I is a global measure of spatial autocorrelation. *Global* here refers to the fact that it computes a single measure for the entire study area. There is a very strong assumption here that the entire area is more or less homogeneous. For those cases in which this does not hold, Anselin [4] proposed a *local* measure based on Moran's I to account for spatial heterogeneity. Those measures are collectively known as LISA, Local Indicators of Spatial Association.

```
plot(pollut_subset$DS_PM_pred,pollut_lag,
    col=alpha("steelblue", 0.5), pch=16, cex=0.4,
    xlab="PM2.5" , ylab="Spatial lag", cex.lab=0.7, cex.axis=0.7)
M1 <- lm(pollut_lag ~ pollut_subset$DS_PM_pred)

abline(M1, col="cyan")
```

This line of code computes the value of the Moran's I statistic

```
# # Moran's I
I <- moran(pollut_subset$DS_PM_pred, pollut_lw, length(pollut_nb), Szero(
     pollut_lw), zero.policy = TRUE)[1]

I
```

```
## $I
## [1] 0.9983595
```

The following block of code performs the Moran's test under both randomization and normality assumptions. The randomization assumption—also known as permutation or conditional randomization [30]—assumes that the observed values are simply a random sample from a population with an unknown distribution. The null hypothesis states that the observed data values could be assigned to any location with equal probability. In practice, it is very common to use the randomization assumption because normality may be too strong a restriction without knowing the underlying distribution. In this case, the results are almost identical for all three indicators (the statistic itself, expectation, and variance).

```
# Moran Test
moran.test(pollut_subset$DS_PM_pred, pollut_lw, alternative="greater")

moran.test(pollut_subset$DS_PM_pred, pollut_lw, alternative="greater",
     randomisation = FALSE)
```

```
##
## Moran I test under randomisation
##
## data: pollut_subset$DS_PM_pred
```

```
## weights: pollut_lw
##
## Moran I statistic standard deviate = 135.36, p-value < 0.0000000000000022
## alternative hypothesis: greater
## sample estimates:
## Moran I statistic Expectation Variance
## 0.99835949638 -0.00012465719 0.00005441368
##
##
## Moran I test under normality
##
## data: pollut_subset$DS_PM_pred
## weights: pollut_lw
##
## Moran I statistic standard deviate = 135.37, p-value < 0.0000000000000022
## alternative hypothesis: greater
## sample estimates:
## Moran I statistic Expectation Variance
## 0.9983594964 -0.0001246572 0.0000544033
```

Similar to Pearson's correlation coefficient, the bounds of the Moran statistic are `-1` and `1`. Values closer to `-1` correspond to a situation in which the values of the variable of interest are highly dispersed. Conversely, values closer to `1` indicate that similar values are highly clustered. Both the scatterplot of the $PM_{2.5}$ variable versus the lagged neighbors, and the value of Moran's I indicate that census tracts with high (or low) values of particulate pollution will be surrounded by census tracts with similarly high (or low) values of particulate pollution. Figure 8.10 shows this relationship.

One of the attractive features of Moran's I algorithm is that it computes very quickly. But this computational efficiency comes with a price. If polygons are distributed in a highly irregular fashion, the values would be potentially biased and unstable. As in many other statistical approaches, we attempt to resolve this shortcoming by computing those statistics within a simulation with thousands of runs, usually based on Monte Carlo (MC) methods. The function `moran.mc` does exactly that. Below, we compute Moran's I with a MC component with 500 permutations

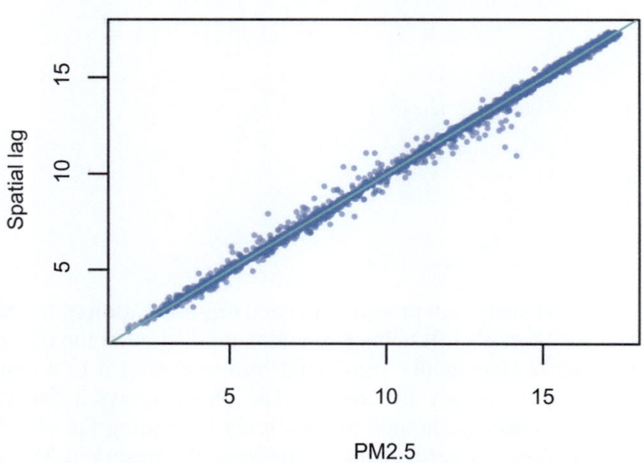

Fig. 8.10 PM2.5 values and spatial lag and regression line

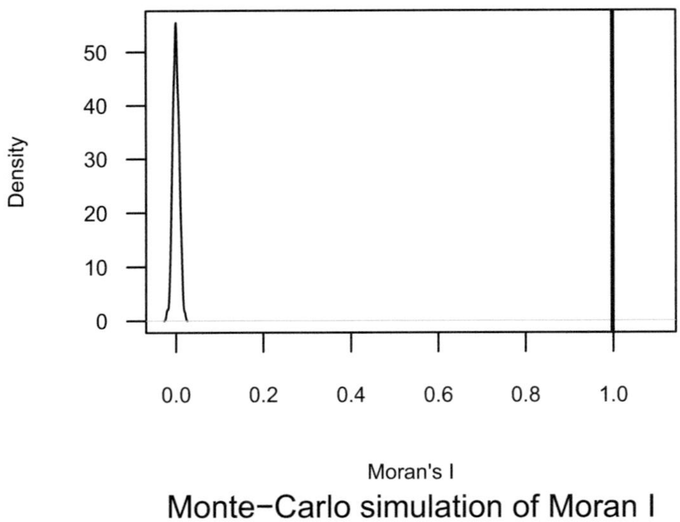

Fig. 8.11 Permutation of Outcomes–Density Plot

and plot the density of permutations in Fig. 8.11. Other aspects of Moran's I are discussed in Li, Calder, and Cressie [31], where the authors offer an alternative to Moran's original formulation.

```
# Run the MC simulation version of the Moran's I test
Moran_mc <- moran.mc(pollut_subset$DS_PM_pred, pollut_lw, nsim=500, alternative
    ="greater")

# Display the resulting statistics
Moran_mc

# Plot the results
plot(Moran_mc, xlab="Moran's I", ylab="Density", main="",
    cex.lab=0.7, cex.axis=0.7, las=1)
```

```
##
## Monte-Carlo simulation of Moran I
##
## data: pollut_subset$DS_PM_pred
## weights: pollut_lw
## number of simulations + 1: 501
##
## statistic = 0.99836, observed rank = 501, p-value = 0.001996
## alternative hypothesis: greater
```

The `moran.mc()` density plot provides a visual representation of the Monte Carlo permutation test for Moran's I. It helps you understand whether the observed spatial pattern in your data is statistically significant by comparing it to a distribution of patterns that could arise purely by chance. The plot displays a density curve of simulated Moran's I values generated by randomly permuting the observed spatial data values. It also shows a vertical line to indicate the observed Moran's I value from your actual data (without permutation). If your observed Moran's I stands out from this random distribution, you have evidence of spatial autocorrelation. This is the case in the plot above. The vertical line at almost 1.0 (0.9983) is far away from the simulated distribution, which is centered at zero.

We close the section on Moran's I by confirming that all measures, the spatial lags, "I", the "I test", and the Monte Carlo simulation, point toward the same conclusion: Areas with high (low) concentration of particular matter pollution neighbor other areas of high (low) pollution.

8.7.3 Geary's C

This measure of spatial autocorrelation, also known as Geary's contiguity ratio (or simply, Geary's ratio), is similar to Moran's I, but it may pick up autocorrelation in cases in which Moran's I may not. This is due to its formulation, which uses squared distances instead of the standardized spatial covariance in the case of Moran's I. For additional details and discussion, see Cliff and Ord [13], Bivand, Pebesma, and Gómez-Rubio [26], Anselin [4], and Haining [6].

The main parameters of the `geary.test` function (in the `spdep` package) are the vector containing the variable of interest, and the weight matrix (`pollut_lw`). In the code block below, we generate Geary's C, both under randomization and under normality

```
# randomization
geary.test(pollut_subset$DS_PM_pred, pollut_lw)
# normality
geary.test(pollut_subset$DS_PM_pred, pollut_lw, randomisation = FALSE)
```

```
##
## Geary C test under randomisation
##
## data: pollut_subset$DS_PM_pred
## weights: pollut_lw
##
## Geary C statistic standard deviate = 132.24, p-value < 0.00000000000000022
## alternative hypothesis: Expectation greater than statistic
## sample estimates:
## Geary C statistic Expectation Variance
## 0.0019338732 1.0000000000 0.0000569635
##
##
## Geary C test under normality
##
## data: pollut_subset$DS_PM_pred
## weights: pollut_lw
##
## Geary C statistic standard deviate = 123.49, p-value < 0.00000000000000022
## alternative hypothesis: Expectation greater than statistic
## sample estimates:
## Geary C statistic Expectation Variance
## 0.00193387321 1.00000000000 0.00006532618
```

We see that the results under both assumptions are very similar, as only the standard deviate and the variance are different. Also, in both cases, the "C" statistic is very small (in the range of 0.002), indicating positive autocorrelation. That is, the findings using Geary's C are identical to those found using Moran's I. The result of the R functions for Moran's I and Geary's C report the *standard deviate*, that is, the differences from the assumed values under the null hypothesis of no spatial autocorrelation, and the corresponding significance. In the example of air pollution from particulate matter, the clustering effects are very strong and unambiguous about their spatial distribution. As a result, Moran's I and Geary's C show identical results.

Anselin [4] and Anselin [32] present the similarities and differences between Moran's I and Geary's C. Like in the case of Moran's I, there are *local* and *global* versions of Geary's C. While the global Moran's I has limits of +1 and −1 (making it as intuitive as Pearson's correlation coefficient), Geary's C has limits of zero and some undefined value larger than +1. Geary's C is centered at +1 (not zero) for the expected value under the null hypothesis of no spatial autocorrelation. This makes it unusual and less intuitive.

In the case of the local Geary's C, the differences between it and Moran's I (as well as Getis-Ord G, as discussed in the next section) are more pronounced. Positive values of Geary's C correspond to "hot spots" (that is, clusters of high values), and negative values correspond to "cold spots" (clusters of low values). This makes it even less intuitive when compared with other measures. There is, however, a certain usefulness in Geary's C. It can help identify unusual locations, a sort of outlier, if you will, worthy of further exploration, even if the source of that location's uniqueness is not detectable directly from the value of Geary's C.

8.7.4 Getis-Ord G

Why do we need a third measure of spatial autocorrelation? First, let's remember that spatial autocorrelation refers to the degree to which the values of a variable at nearby locations are similar or dissimilar. The Getis-Ord G (G_i^*, pronounced G-i-star), is a spatial autocorrelation measure that stands out from others due to its specific focus on identifying hot spots and cold spots—areas where high or low values of a variable cluster spatially [7].

```
# Test for global G statistic of PM2.5
globalG.test(pollut_subset$DS_PM_pred, pollut_lw)

# Local Gi Test
# Test for local spatial autocorrelation (hotspots)
# Identify neighbors, create weights, calculate spatial lag
pollut_subset <- st_as_sf(pollut_subset)

# neighbors share border/vertex
# row-standardized weights
# calculate spatial lag of DS_PM_pred
pollut_nbs <- pollut_subset |>
  mutate(
    nb = st_contiguity(geometry),
    wt = st_weights(nb),
    pollut_lag = st_lag(DS_PM_pred, nb, wt)
  )
```

```
##
## Getis-Ord global G statistic
##
## data: pollut_subset$DS_PM_pred
## weights: pollut_lw
##
```

```
## standard deviate = 77.181, p-value < 0.0000000000000022
## alternative hypothesis: greater
## sample estimates:
## Global G statistic Expectation Variance
## 0.00014888659015286230 0.00012465719272000477 0.0000000000009855261
```

Getis-Ord G calculates a statistic for each individual feature in a dataset (in our example, census tract polygons). It tells you whether that feature and its neighbors have values that are significantly higher or lower than the average for the entire study area. In contrast, Moran's I (especially the global version) indicates the overall presence and strength of spatial autocorrelation (clustered or dispersed) but doesn't differentiate between clusters of high values and clusters of low values. A positive Moran's I indicates clustering, but you need to visually inspect the data or use a local Moran's I to see if it's high-high or low-low clustering.

The map in Fig. 8.12 shows the Getis-Ord G values for all census tracts in California. A high positive G indicates a statistically significant hot spot, meaning a cluster of high values. The larger the positive, the more intense the clustering of high values.

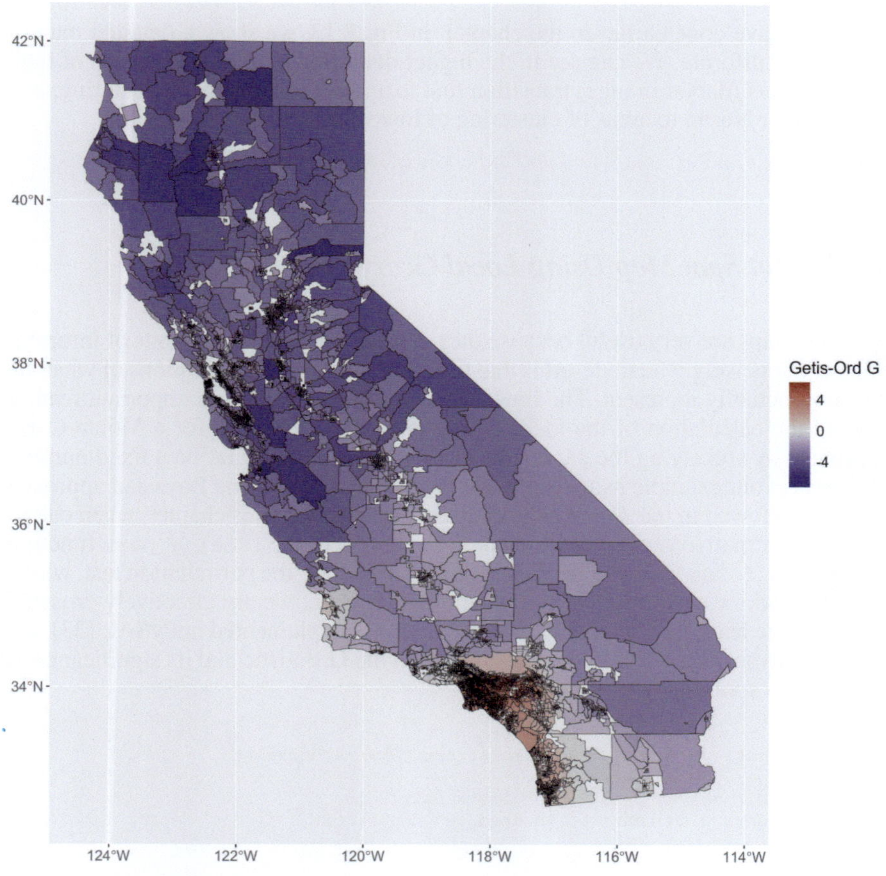

Fig. 8.12 Getis-Ord Gi* Hot Spot Map

A low negative G indicates a statistically significant cold spot, meaning a cluster of low values. The smaller (more negative) the value, the more intense the clustering of low values. A value near zero indicates no apparent clustering.

```
# Calculate the Gi using localG_perm
pollut_hot_spots9 <- pollut_nbs |>
  mutate(G1 = localG_perm(pollut_subset$DS_PM_pred, pollut_lw, nsim = 499)) |>
  unnest(G1)
```

```
pp4 <- ggplot(pollut_hot_spots9) +
  geom_sf(aes(fill = as.numeric(G1)), color = "black", lwd = 0.15) +
  scale_fill_gradient2(low = "blue4", mid = "lightgray", high = "red4") +
  labs(
    fill = "Getis-Ord G" ) +
  theme(
    axis.title = element_blank(),
    axis.text = element_text(size = 5),
    panel.grid = element_line(color = "white", linewidth = 0.25),
    legend.text = element_text(size = 6),
    legend.title = element_text(size = 7))

pp4
```

As we have done earlier in the chapter, in Fig. 8.13, we show a detailed map of Southern California. We can see in the higher definition map how the areas of high value clusters (darker orange) transition first to an area of no evident clustering, and then, further North, to areas of clustering of low values (shades of purple).

```
pp4 + coord_sf(xlim = c(-119,-116.5), ylim=c(32,34.5), expand=FALSE)
```

8.7.5 *Hot Spot Map Using Local Geary's Statistic*

Hot spot maps are very useful because they communicate large amounts of information very concisely. The trade-off is that they require careful explanations of what the hot spots actually represent. The function `local_g_perm` adds an important feature beyond the calculation of the *local Geary statistic* as it allows for a Monte-Carlo approach, by specifying the parameter `nsim` to run the computation a fixed number of times (in our example nsim=500). In essence, this is the same Bayesian approach we have followed in the *Time Series* chapter, and earlier in this chapter, when calculating an alternative version of Moran's I. The output of the `local_g_perm` function yields the `p_folded_sum` value This is the expression of the permutation test, which is performed for each iteration of the simulation. Here, we are effectively crossing into machine learning approaches. The methods are implemented in PySAL [33], the spatial analysis library for Python. We use both the G statistic and its significance to map the hot spots.

```
# Calculate the Gi using local_g_perm
# nsim = number of Monte Carlo simulations (999 is default)

# The new 'Gi' column itself contains a dataframe
# We can't work with that, so we need to 'unnest' it

pollut_hot_spots2 <- pollut_nbs |>
  mutate( Gi = local_g_perm(DS_PM_pred, nb, wt, nsim = 500)) |>
  unnest(Gi)
```

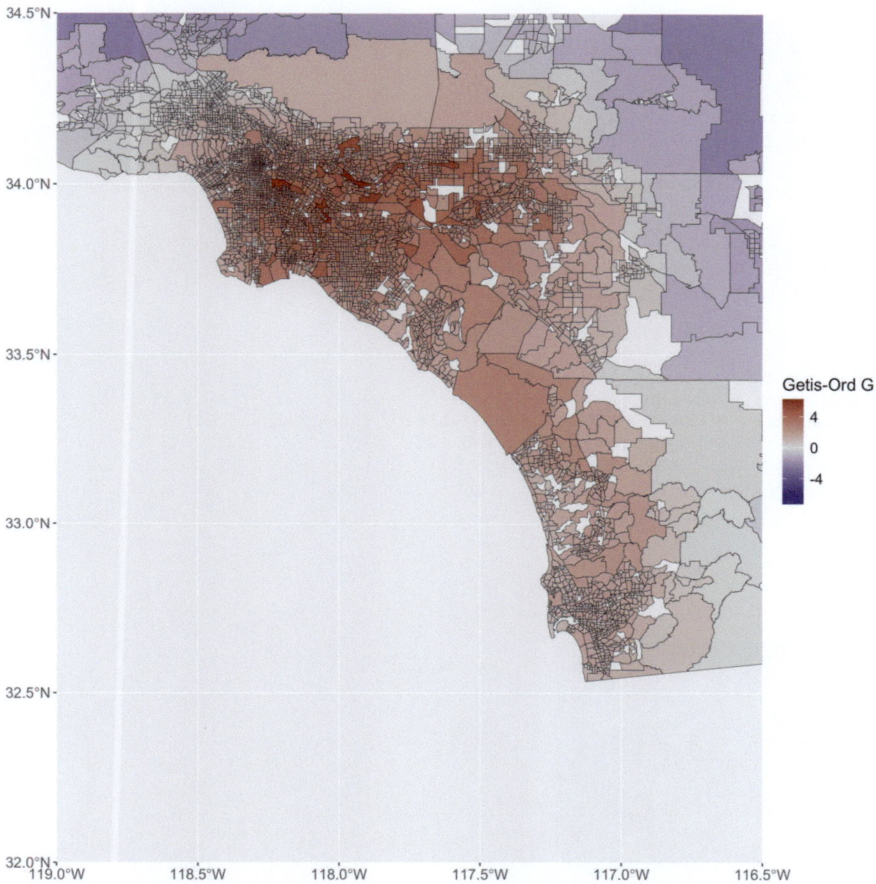

Fig. 8.13 Getis-Ord Gi* Hot Spot Map of Southern California

The map in Fig. 8.14 shows the pollution hot spots for the whole state, and in Fig. 8.15, we can see the details for Southern California.

```
# Create a new data frame called 'pollut_hot_spots"
# with the columns 'gi' and 'p_folded_sim"
# 'p_folded_sim' is the p-value of a folded permutation test [PySAL]

# Add a new column called "classification"; convert 'classification' into a
# factor for easier plotting; and visualize the results with ggplot2

# Classify based on the following criteria:
pp5 <- pollut_hot_spots2 |>
  select(gi, p_folded_sim) |>
  mutate(
    classification = case_when(
      gi > 0 & p_folded_sim <= 0.01 ~ "Very high",
      gi > 0 & p_folded_sim <= 0.05 ~ "High",
      gi > 0 & p_folded_sim <= 0.1 ~ "Somewhat high",
      gi < 0 & p_folded_sim <= 0.01 ~ "Very Low",
      gi < 0 & p_folded_sim <= 0.05 ~ "Low",
```

```
    gi < 0 & p_folded_sim <= 0.1 ~ "Somewhat Low",
    TRUE ~ "Non-significant"
   ),
   classification = factor(
    classification,
    levels = c("Very high", "High", "Somewhat high",
            "Non-significant",
            "Somewhat Low", "Low", "Very Low")
   )
  ) |>
  ggplot(aes(fill = classification)) +
  geom_sf(color = "black", lwd = 0.075) +
  scale_fill_brewer(type = "div", palette = "Spectral") +
  labs(
   fill = "Hot Spot Classification") +
  theme(
   axis.title = element_blank(),
   axis.text = element_text(size = 5),
   panel.grid = element_line(color = "white", linewidth = 0.25),
   legend.text = element_text(size = 6),
   legend.title = element_text(size = 7))

pp5
```

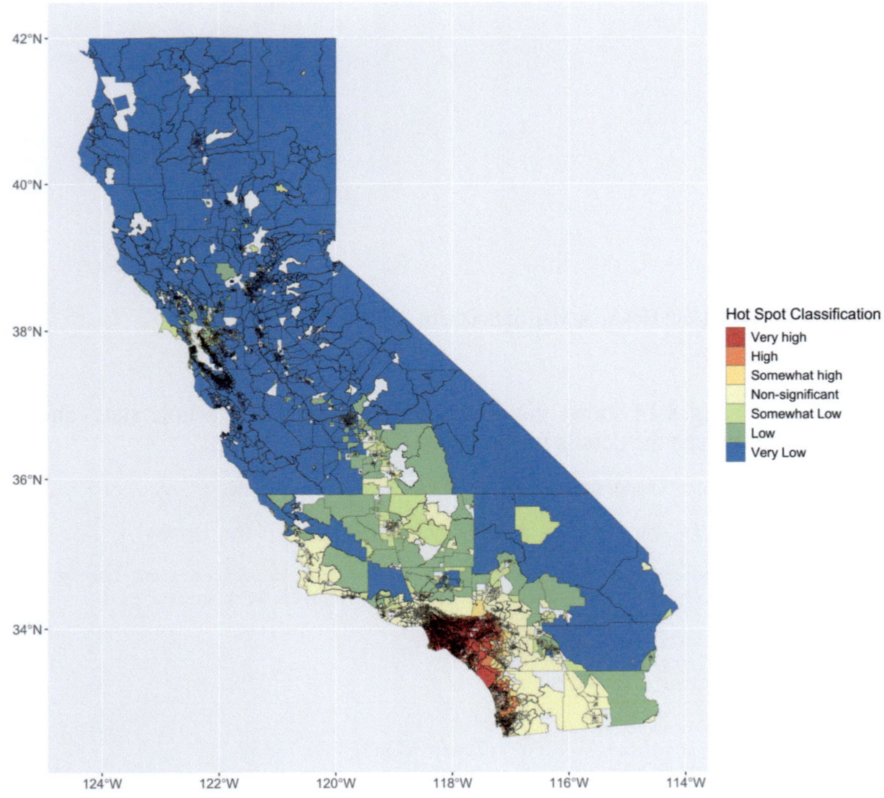

Fig. 8.14 Hot spots based on Local Geary's statistic

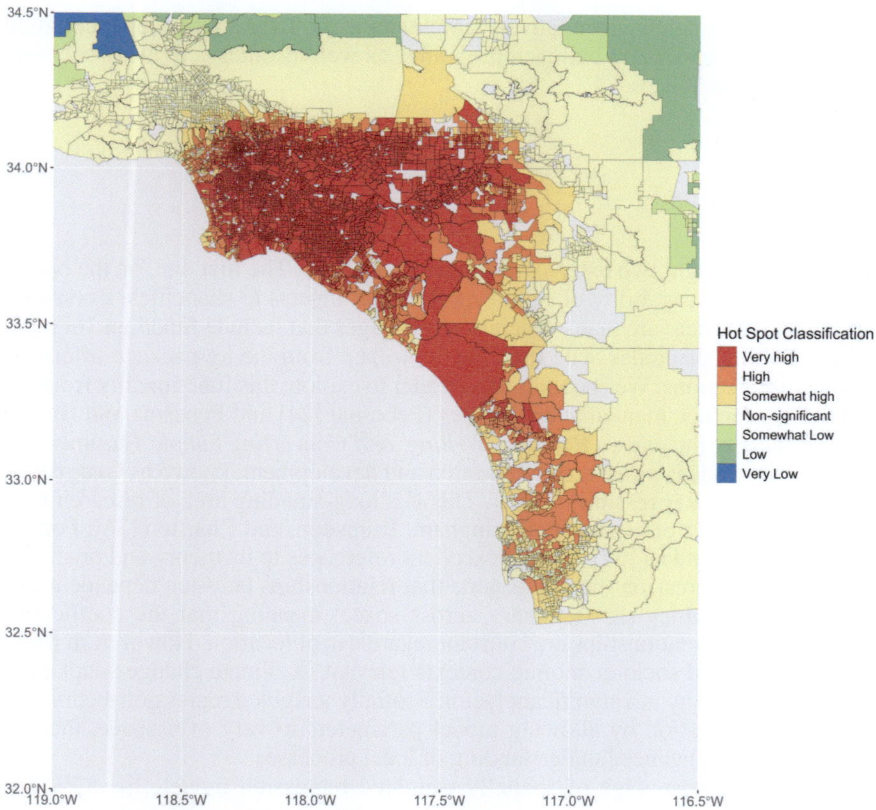

Fig. 8.15 Hot spots in Southern California

The code chunk we used to generate the hot spot map is a modified version of Parry [34] on hot spot analysis. A very similar implementation for a different case is found in Leary [35]. We acknowledge these authors, and others, as we found significant inspiration from their contributions.

We can now generate a map for Southern California, so we can see more details about how hot spots are distributed.

```
pp5 + coord_sf(xlim = c(-119,-116.5), ylim=c(32,34.5), expand=FALSE)
```

The previous maps (Figs. 8.14 and 8.15) show the distribution of similar levels of air pollution at the census tract level. This is a very useful application of spatial autocorrelation measures, as it shows the similarities of a given attribute across spatial units. In the California example, we knew the levels of air pollution from the map in Fig. 8.3. But now we also know that areas of high levels of pollution also neighbor other areas of high levels of pollution in Southern California. Similarly, for the rest of the state, areas of low or very low levels are near other areas of low or very low levels of pollution.

In general, hot spot maps are not the ultimate analytical tool. They are, however, efficiently used as part of sophisticated communication plans as they convey large amounts of information with relatively simple elements. Their application is common in public engagement and outreach programs, as well as other activities for broad, non-technical audiences.

8.8 Extensions

We want to highlight two extensions of spatial analysis. The first one, on the operational side, is the capability of exporting R spatial objects to shapefiles, a common format in GIS. There are many specialized packages that include functions for reading and writing shapefiles. The function `st_write` from the `sf` package offers the broadest applications. We encourage the reader to explore this functionality from the package's reference manual and vignettes (Pebesma [24] and Pebesma and Bivand [10]), especially the one on *Reading, Writing, and Converting Simple Features*.

The second one is a vast area of research and development, known as **Geographically Weighted Regression** (GWR). This is a long-standing area of research with more than 25 years of history. Fotheringham, Brunsdon, and Charlton [36]; Fotheringham, Oshan, and Li [37] provide excellent references to its theory and practice.

Traditional regression models assume that relationships between dependent and independent variables are stationary across space, meaning that the coefficients describing these relationships are constant regardless of location. However, in many environmental and socio-economic contexts relevant to climate change adaptation, spatial heterogeneity is a significant factor. Spatially weighted regression techniques address this limitation by allowing model parameters to vary over space, thereby providing a more nuanced understanding of local processes.

The primary advantage of spatially weighted regression models like GWR is their ability to identify and visualize spatial non-stationarity in relationships among variables. For practitioners in climate change adaptation, this allows for the identification of "hotspots" where certain drivers of vulnerability or resilience operate differently, or where the effectiveness of adaptation interventions might vary over space For instance, the relationship between extreme heat exposure and health outcomes might be stronger in neighborhoods with older housing stock and less green space, a spatial variation that a global regression model would obscure. By providing local coefficients, spatially weighted regression offers valuable insights for targeted and place-based adaptation strategies, moving beyond a "one-size-fits-all" approach.

Implementing Geographically Weighted Regression (GWR) in R provides a powerful and flexible environment for spatial analysis. Several packages facilitate GWR, with `spgwr` [38] and `GWmodel` [39] being the most prominent. These packages offer functions for the GWR model fitting itself, along with tools for visualizing and interpreting the spatially varying coefficients.

By leveraging these R packages, researchers and practitioners can effectively implement and interpret GWR models, providing crucial spatially explicit insights for climate change adaptation planning and policy development.

Fortunately, there is a growing literature on geospatial processing in R. Multiple packages are already available, and developers continuously make significant contributions. Many books have also made a significant impact in making detailed workflows available to researchers and practitioners. One such example is Lovelace, Nowosad, and Muenchow [40] and its companion website.

8.9 Closing Comments

In this chapter, we demonstrated the value of spatial autocorrelation diagnostics in the context of environmental data. Using $PM_{2.5}$ pollution estimates across California census tracts, we applied three widely used measures—Moran's I, Geary's C, and Getis-Ord G—to evaluate the extent and patterns of spatial clustering. Each of these measures offers a unique lens: Moran's I provides an efficient global summary; Geary's C captures subtle local variations; and Getis-Ord Gi* pinpoints hot and cold spots. Together, they offer a comprehensive spatial analytical toolkit that is essential for climate adaptation planning. Understanding where environmental burdens concentrate can inform more equitable, targeted policy responses and mitigation strategies.

8.10 Workflows in Spatial Analysis

Workflows for spatial analysis are a function of the purpose of the analysis. There are so many different aspects that we cannot describe all of them in a book that has a general overview approach. There are, however, some commonalities, and we will focus on those.

1. Data preparation is a fundamental component, especially because we can regularly combine geographic or spatial data with tabular data.
    a. Make sure that the geographic definitions are congruent between data sources (i.e., tabular data at the county level to match with county maps)
    b. Look for discrepancies in the data, such as empty polygons, a mismatch between spatial and geographic data, and misidentification of spatial units (especially when compiling data from different sources)
    c. If there will be a combination of vector data sources, or vector and raster data, make sure that the scales, projection methods, and datum are congruent.
2. Generate a set of relevant descriptive spatial statistics
3. If applicable, split the data into *training* and *testing* sets for the purpose of validating the forecasting algorithm.
4. Compute appropriate performance metrics to assess the model quality.

References

1. Moran, Patrick Alfred Pierce. 1950. Notes on continuous stochastic phenomena. *Biometrika* 37(1-2): 17–23. ISSN: 0006–3444. https://doi.org/10.1093/biomet/37.1-2.17.
2. Geary, R. C. 1954. The contiguity ratio and statistical mapping. *The Incorporated Statistician* 5(3): 115–146. ISSN: 14669404. https://doi.org/10.2307/2986645.
3. Tobler, W. R. 1970. A computer movie simulating urban growth in the Detroit region. *Economic Geography* 46: 234–240. ISSN: 00130095, 19448287. https://doi.org/10.2307/143141.
4. Anselin, Luc. 1995. Local indicators of spatial association–LISA. *Geographical Analysis* 27(2): 93–115. https://doi.org/10.1111/j.1538-4632.1995.tb00338.x.
5. Agarwal, Mohit, et al. 2024. General Geospatial Inference with a Population Dynamics Foundation Model. In arXiv: 2411.07207 `[cs.LG]`. https://arxiv.org/abs/2411.07207.
6. Haining, Robert. 2003. *Spatial data analysis: Theory and practice*. Cambridge University Press. ISBN: 0-521-77437-3.
7. Getis, Arthur, and Ord, J. K. 1992. The analysis of spatial association by use of distance statistics. *Geographical Analysis* 24(3): 189–206. https://doi.org/10.1111/j.1538-4632.1992.tb00261.x.
8. Getis, Arthur, and Jared Aldstadt. 2004. Constructing the spatial weights matrix using a local statistic. *Geographical Analysis* 36(2): 90–104. https://doi.org/10.1111/j.1538-4632.2004.tb01127.x.
9. Fischer, Manfred M., and Arthur Getis (eds.). 2010. *Handbook of applied spatial analysis*. Springer Books 978–3-642-03647-7. Springer. ISBN: ARRAY(0x70cc38a8). https://link.springer.com/book/10.1007/978-3-642-03647-7. https://doi.org/10.1007/978-3-642-03647-7.
10. Pebesma, Edzer, and Roger Bivand. 2023. *Spatial data science with applications in R*. Chapman & Hall. https://r-spatial.org/book/.
11. Simonovska, Rozeta. 2024. *SDPDmod: Spatial dynamic panel data modeling*. https://CRAN.R-project.org/package=SDPDmod.
12. Moran, Patrick Alfred Pierce. 1948. The Interpretation of Statistical Maps. *Journal of the Royal Statistical Society Series B-Methodological* 10: 243–251. https://doi.org/10.1111/j.2517-6161.1948.tb00012.x.
13. Cliff, A. D., and J. K. Ord. 1973. *Spatial autocorrelation*. Pion. ISBN: 9780850860375.
14. California Air Resources Board. 2025. Inhalable Particulate Matter and Health (PM2.5 and PM10). https://ww2.arb.ca.gov/resources/inhalable-particulate-matter-and-health. Accessed: June 2025.
15. Arden Pope III, C., and Douglas W. Dockery. 2006. Health effects of fine particulate air pollution: Lines that connect. *Journal of the Air & Waste Management Association* 56(6): 709–742. https://doi.org/10.1080/10473289.2006.10464485.
16. World Health Organization. 2021. WHO Global Air Quality Guidelines: Particulate matter (PM2.5 and PM10), ozone, nitrogen dioxide, sulfur dioxide and carbon monoxide. https://apps.who.int/iris/handle/10665/345329.
17. Seinfeld, J.H., and S.N. Pandis. 2016. *Atmospheric chemistry and physics: From air pollution to climate change*. Wiley.
18. Reid, Colleen E., et al. 2016. Critical review of health impacts of wildfire smoke exposure. *Environmental Health Perspectives* 124(9): 1334–1343. https://doi.org/10.1289/ehp.1409277.
19. World Meteorological Organization. 2023. State of the Global Climate 2023. https://wmo.int/publication-series/state-of-global-climate-2023.
20. Gil-Alana, Luis A., and Manuel Monge. 2020. Global CO emissions and global temperatures: Are they related. *International Journal of Climatology* 40(15): 6603–6611. https://doi.org/10.1002/joc.6601.
21. Reid, Colleen E., et al. 2021. Daily PM2.5 concentration estimates by county, ZIP code, and census tract in 11 western states 2008–2018. *Scientific Data* 8(1): 112. ISSN: 2052–4463. https://doi.org/10.1038/s41597-021-00891-1.
22. Di, Qian, et al. 2019. An ensemble-based model of PM2.5 concentration across the contiguous United States with high spatiotemporal resolution. *Environment International* 130: 104909. ISSN: 0160–4120. https://doi.org/10.1016/j.envint.2019.104909.
23. Walker, Kyle. 2016. tigris: An R package to access and work with geographic data from the US Census Bureau. *The R Journal* 8(2): 231–242. https://doi.org/10.32614/RJ-2016-043. https://doi.org/10.32614/RJ-2016-043.

24. Pebesma, Edzer. 2018. Simple features for R: Standardized support for spatial vector data. *The R Journal* 10(1): 439–446. https://doi.org/10.32614/RJ-2018-009. https://doi.org/10.32614/RJ-2018-009.
25. Morello-Frosch, Rachel, et al. 2011. The climate gap: Inequalities in how climate change hurts Americans & how to close the gap. *Race, Poverty & The Environment* 18(2): 19–22.
26. Bivand, Roger, Edzer Pebesma, and Virgilio Gómez-Rubio. 2013. *Applied spatial data analysis with R*, 2nd ed. NY: Springer. https://doi.org/10.1007/978-1-4614-7618-4.
27. Anselin, Luc. 1988. *Spatial econometrics: Methods and models*, vol. 4. Springer.
28. LeSage, J., and R. K. Pace. 2009. *Introduction to Spatial Econometrics*. Chapman and Hall/CRC. https://doi.org/10.1201/9781420064254.
29. Bivand, Roger. 2022. R packages for analyzing spatial data: A comparative case study with areal data. *Geographical Analysis* 54(3): 488–518. https://doi.org/10.1111/gean.12319. eprint: https://onlinelibrary.wiley.com/doi/pdf/10.1111/gean.12319. https://onlinelibrary.wiley.com/doi/abs/10.1111/gean.12319.
30. Bivand, Roger, and David Wong. 2018. Comparing implementations of global and local indicators of spatial association. *TEST* 27(3): 716–748. https://doi.org/10.1007/s11749-018-0599-x. https://doi.org/10.1007/s11749-018-0599-x.
31. Li, Hongfei, Catherine A. Calder, and Noel Cressie. 2007. Beyond Moran's I: Testing for spatial dependence based on the spatial autoregressive model. *Geographical Analysis* 39(4): 357–375. https://doi.org/10.1111/j.1538-4632.2007.00708.x.
32. Anselin, Luc. 2019. A local indicator of multivariate spatial association: Extending Geary's c. *Geographical Analysis* 51(2): 133–150. https://doi.org/10.1111/gean.12164.
33. Rey, Sergio J., and Luc Anselin. 2007. PySAL: A Python library of spatial analytical methods. *The Review of Regional Studies* 37(1): 5–27.
34. Parry, Josiah. 2022. Hotspot Analysis. https://github.com/JosiahParry/youtube-tutorials/tree/main/hot-spot-analysis
35. Leary, Heatherlee. 2023. R tutorial: Hotspot analysis using Getis-Ord "Gi". https://rpubs.com/heatherleeleary/hotspot_getisOrd_tut.
36. Fotheringham, A.S., C. Brunsdon, and M. Charlton. 2002. *Geographically Weighted Regression: The Analysis of Spatially Varying Relationships*. Wiley. ISBN: 9780470855256.
37. Fotheringham, A.S., T.M. Oshan, and Z. Li. 2023. *Multiscale geographically weighted regression: Theory and practice*. CRC Press. ISBN: 9781000989700.
38. Bivand, Roger, and Danlin Yu. 2024. spgwr: Geographically weighted regression. R package version 0.6-37. https://CRAN.R-project.org/package=spgwr.
39. Lu, Binbin, et al. 2014. The GWmodel R package: Further topics for exploring spatial heterogeneity using geographically weighted models. *Geo-spatial Information Science* 17(2): 85–101. https://doi.org/10.1080/10095020.2014.917453.
40. Lovelace, Robin, Jakub Nowosad, and James Muenchow. 2025. *Geocomputation with R*. 2nd. The R Series. Boca Raton, FL: Chapman and Hall/CRC. https://r.geocompx.org/.

Part V
Relationships in Survey Data

Chapter 9
Introduction to Non-parametric Methods

In the introductory chapter of this book, we presented a typology of data structures and levels of measurement. The two principal categories—**metric** and **categorical** data—were further subdivided: metric data into *ratio* and *interval* levels, and categorical data into *ordinal* and *nominal* levels. At that stage, we did not elaborate on the implications that these data types have for the selection of appropriate statistical methods. However, understanding these implications is critical for rigorous empirical analysis.

Broadly speaking, metric data are typically analyzed using *parametric* statistical tests, while categorical data are more often examined using *non-parametric* methods. The key distinction between these two families of tests lies in their underlying assumptions about data distribution. A rather restrictive set of assumptions is mandatory in parametric tests. The most salient ones are that the data follow a specific distribution, homoscedasticity, and interval-level measurement. Non-parametric tests, by contrast, are more flexible and impose fewer distributional requirements. Indeed, they are sometimes referred to as *distribution-free* methods because they can be applied even when the data do not conform to any known or hypothesized probability distribution.

This distinction is particularly relevant in the field of policy analysis, where much of the empirical evidence is derived from surveys, opinion polls, and qualitative research. These instruments typically yield large volumes of categorical data. For example, ordinal responses might capture the degree of agreement with a policy proposal, the perceived importance of an issue, or the intensity of concern. Nominal responses, on the other hand, may capture categorical identifications such as political affiliation, demographic characteristics, or binary responses (e.g., yes/no, support/oppose).

Even more sophisticated forms of qualitative data collection—such as text analysis, structured or semi-structured interviews, or ethnographic field notes—ultimately produce categorical data once coding is applied. These coding processes often involve assigning responses or observations into discrete categories, resulting in large matrices of nominal or ordinal data that can be summarized using frequencies or contingency tables. Thus, despite the complexity of the data collection process, the analytical phase frequently relies on the logic and tools of categorical data analysis.

F. DePaolis, *Workflows for Climate Change Adaptation Analytics*,
https://doi.org/10.1007/978-3-032-14004-3_9

Given the prevalence of such data types in policy research, non-parametric techniques become indispensable. While these methods may not yield the same inferential power as parametric alternatives under ideal conditions, they are robust in real-world contexts where distributional assumptions are often violated. This robustness makes non-parametric methods especially valuable in climate adaptation research, where survey-based data are used to understand public sentiment, behavioral adaptations, and community priorities.

For a comprehensive treatment of techniques suited for categorical data analysis, see Agresti [1] and McNabb [2], both of whom provide extensive methodological guidance for working with nominal and ordinal data in applied settings.

9.1 Data Structures

To illustrate the utility of non-parametric methods in policy-relevant analysis, consider the following example. Suppose we wish to test whether there is a statistically significant association between two categorical survey responses. The first item states: *Climate change is caused or accelerated by human activities*, with possible responses: agree, neutral, and disagree. The second item reads: *I would consider participating in a voluntary managed retreat program*, with possible responses: likely, neutral, and unlikely. Both variables are ordinal, and there is no reason to assume that the responses are normally distributed. In such a case, parametric techniques would be inappropriate, and a non-parametric method—such as the Chi-square test for independence or a measure of ordinal association like Kendall's tau—would be more suitable for assessing the relationship between these two variables.

Contingency Tables

We use the term 'contingency table' for a data structure containing the 'joint distribution' of the responses. In this case, the table has three rows and three columns (3 by 3 table). The table below shows such a structure, adding a row at the bottom for the column sums.

```
# READ DATA
opinion_poll <- read_csv("data/raw-non-param.csv", col_names = TRUE)
opinion_poll$Participation <- factor(opinion_poll$Participation, levels = c("
     Likely", "Neutral", "Unlikely"))
opinion_poll$Opinion <- factor(opinion_poll$Opinion, levels = c("Agree", "
     Neutral", "Disagree"))
# Creates contingency table
cont_table <- table(opinion_poll$Opinion, opinion_poll$Participation)

# Calculate expected values
exp_table <- matrix(0, nrow = nrow(cont_table), ncol = ncol(cont_table))

for (i in 1:nrow(cont_table))
  {for (j in 1:ncol(cont_table))
       {exp_table[i, j] <- (rowSums(cont_table)[i] * colSums(cont_table)[j]) /
            sum(cont_table)
  }
}

rownames(exp_table) <- c("Agree", "Neutral", "Disagree")
colnames(exp_table) <- c("Likely", "Neutral", "Unlikely")
```

```
kable(cont_table, caption="Observed distribution", digits = 3,
     format = "latex", booktabs = TRUE, linesep="")|>
     add_header_above(header = c(" " = 1, "Partipation" = 3))|>
     pack_rows("Climate Change", start_row = 1, end_row = 3, bold = F)|>
  kable_styling(latex_options = "HOLD_position", "scale_down")
```

```
kable(exp_table, caption="Expected distribution", digits = 3,
     format = "latex", booktabs = TRUE, linesep="")|>
     add_header_above(header = c(" " = 1, "Partipation" = 3))|>
     pack_rows("Climate Change", start_row = 1, end_row = 3, bold = F)|>
  kable_styling(latex_options = "HOLD_position", "scale_down")
```

9.2 Tests

The most straightforward test we will run is the chi-squared test (χ^2), from the Greek letter χ (*chi*) that resembles the Latin letter 'x'. In one of its forms, the null hypothesis of this test is that the proportions in each row of the contingency table are identical to the totals row. The conceptual null hypothesis states that the responses to the two questions are statistically independent, and the table that represents it is the 'expected values' table (shown below).

Both tables, *observed* and *expected* distributions shown in Tables 9.1 and 9.2, respectively, have identical *marginal distributions*. That is, the row and column totals are the same in both cases. But they have different conditional distributions, that is, the values corresponding to each combination of the two attributes. The two tables will be identical when the responses to the two questions are statistically independent.

As the proportions depart from that identity, the association between the responses to the two questions becomes stronger. The issue is how different those must be before concluding their responses are statistically associated. Sometimes, the association is so strong that the table displays a 'diagonal effect.' For example, the table above displays large values in the cells corresponding to 'Disagree-Unlikely' (55), 'Agree-Likely' (41), and much smaller values elsewhere. In some cases, the contingency table

Table 9.1 Observed distribution

	Participation		
	Likely	Neutral	Unlikely
Climate change			
Agree	41	4	4
Neutral	30	7	16
Disagree	21	2	55

Table 9.2 Expected distribution

	Partipation		
	Likely	Neutral	Unlikely
Climate change			
Agree	25.044	3.539	20.417
Neutral	27.089	3.828	22.083
Disagree	39.867	5.633	32.500

has no strong diagonal effect, or its direction is ambiguous; we must then compute a numeric value to capture the association between the observed and expected values. In R, we can compute both the actual value of the χ^2 statistic and the corresponding p-value.

The equation for the test of independence is:

$$\chi^2 = \sum \frac{(O_{ij} - E_{ij})^2}{E_{ij}} \tag{9.1}$$

where O_{ij} is the observed value and E_{ij} is the expected value. The sub-indices ij refer to the ith row and j^{th} column in the corresponding tables.

```
# Calculate chi-square statistic
chi_sq_result <- chisq.test(cont_table)
# Print the statistic and its p-value
print(round(chi_sq_result$statistic, digits = 2))
print(format(chi_sq_result$p.value, digits = 3, scientific = TRUE))
```

```
## X-squared
## 54.89
## [1] "3.42e-11"
```

The equation for χ^2 in (9.1) provides some hints about the limits of the test. When the expected values are very small, either because the total sample size is small or there are many categories, the test is not reliable. This is so because χ^2 becomes artificially large due to the small value of the denominator, E_{ij}. In such cases, the test is not appropriate, and we must use other tests.

A second test, the *likelihood ratio test*, is also available in R. This test is more sensitive to the presence of small cell counts in the contingency table. The likelihood ratio test is a more conservative test, and it is often used when the χ^2 test is not appropriate. The likelihood ratio test is also known as the 'G-test,' and represented by G^2.

Equation (9.2) shows the formulation for the maximum likelihood measure:

$$G^2 = 2 \sum O_{ij} \log(\frac{O_{ij}}{E_{ij}}) \tag{9.2}$$

where O_{ij} and E_{ij} are the observed and expected values, respectively.

```
# Calculate G-square statistic
GTest(cont_table, correct = "none")
```

```
##
## Log likelihood ratio (G-test) test of independence without correction
##
## data: cont_table
## G = 59.428, X-squared df = 4, p-value = 0.000000000003826
```

The results are almost identical in value to those from the χ^2 test above. This is expected because both tests are very similar, especially for cases as the example here, where the sample size is large.

9.3 Additional Considerations and Other Tests

While not covered in detail in this section, two additional statistical tests merit mention due to their importance in the analysis of categorical data, especially for small datasets: Fisher's exact test [3] and Yates' continuity correction [4].

Fisher's exact test is a non-parametric statistical procedure used to evaluate the association between two categorical variables in a 2×2 contingency table. Unlike the chi-squared test, which relies on large-sample approximations, Fisher's test computes the exact probability of obtaining the observed distribution (or one more extreme) under the null hypothesis of independence. The test is grounded in the hypergeometric distribution and remains valid regardless of how small the expected frequencies may be. An extension of this test to larger contingency tables is known as the Fisher-Freeman-Halton test.

Yates' continuity correction, also known as the Yates-corrected chi-squared test, adjusts the standard χ^2 statistic to better approximate the discrete distribution of the test statistic under the null hypothesis, particularly for 2×2 tables with small sample sizes. This correction subtracts 0.5 from the absolute difference between each observed and expected frequency before squaring, thereby reducing the risk of Type I error in small-sample contexts.

These methods provide more reliable inference than the standard χ^2 test when assumptions related to minimum expected cell counts are violated. Most statistical software, R included, will alert the user when the expected frequencies are small. In general, R issues a warning, only alerting the user that the value of chi-squared may be incorrect. Others, such as SPSS, will report the number of cells for which the expected frequency is below 5. In those cases, applying a correction becomes necessary.

Analogous Non-parametric Tests for Categorical Data

A useful framework for understanding non-parametric statistics is through their conceptual equivalence to well-known parametric tests. For instance, the chi-squared test, discussed previously, can be viewed as a non-parametric analog to Pearson's correlation coefficient, in that both assess the association between two variables—though the former applies to categorical data, and the latter to continuous variables. This type of classification is very common in introductory-level statistics textbooks, and we encourage the reader to explore the multiple options available, such as Sheskin [5] and McNabb [2].

Several other non-parametric statistics serve as counterparts to parametric correlation and association measures:

- **Spearman's rank correlation coefficient** (ρ): This statistic evaluates the monotonic relationship between two ranked variables. Like Pearson's r, Spearman's ρ ranges from -1 to 1, indicating the direction and strength of the association, but it does not assume linearity or normally distributed data.
- **Kendall's Tau** (τ): Designed for ordinal data, Kendall's Tau measures the strength and direction of association between two variables based on the concordance and

discordance of paired observations.[1] There are two variants: Tau-b is used when the number of rows and columns in the contingency table is equal, while Tau-c is appropriate when they differ [6].

- **Cramér's V**: An extension of the χ^2 test for larger contingency tables, Cramér's V measures the strength of association between two nominal variables, with values ranging from 0 (no association) to 1 (perfect association).
- **Phi coefficient** (ϕ): Conceptually similar to Pearson's correlation coefficient, the Phi coefficient quantifies the association between two binary (dichotomous) variables in a 2×2 table. It is effectively a special case of Cramér's V for 2×2 matrices.

These non-parametric association measures allow for robust analysis of categorical and ordinal data where parametric assumptions do not hold.

Rank-Based Alternatives to Parametric Tests

The equivalence between parametric and non-parametric methods extends to hypothesis testing for group comparisons. In particular:

- The Kruskal–Wallis H test serves as a non-parametric alternative to the one-way Analysis of Variance (ANOVA) when comparing more than two independent groups. Like ANOVA, it tests the null hypothesis that the group medians are equal, but without requiring assumptions of normality or homogeneity of variances.
- For two independent samples, the Mann–Whitney U test (not mentioned earlier but worth noting here) is the non-parametric analog to the independent samples t-test.

References

1. Agresti, Alan. 2007. *An introduction to categorical data analysis*, 2nd ed. Wiley-Blackwell. https://onlinelibrary.wiley.com/doi/book/10.1002/0470114754.
2. McNabb, David E. 2018. *Research methods in public administration and nonprofit management*. Fourth. Armonk, N.Y.: Routledge.
3. Agresti, Alan. 1992. A survey of exact inference for contingency tables. *Statistical Science* 7(1): 131–153. https://doi.org/10.1214/ss/1177011454.
4. Stefanescu, Catalina, Vance W. Berger, and Scott L. Hershberger. 2014. Yates' correction. In *Wiley StatsRef: Statistics Reference Online*. Wiley. https://doi.org/10.1002/9781118445112.stat06152.
5. Sheskin, David J. 2011. *Handbook of parametric and nonparametric statistical procedures*, 5th. Chapman and Hall/CRC. ISBN: 978-1439858011. https://doi.org/10.1201/9780429186196.
6. Agresti, Alan, and Barbara Finlay. 2014. *Statistical methods for the social sciences*. eng. 4th ed. Harlow, England: Pearson. ISBN: 1-292-03489-0.

[1] A pair of observations is concordant if their relative ordering is the same for both variables. In discordant pairs the relative order is opposite.

Summary of Workflows

Descriptives and Visualization

Creating effective descriptions of datasets and complementing them with impactful visualizations is, first and foremost, a function of understanding the type of data we are analyzing. The following classification sums up the most common components of descriptive statistics by data type

Metric data	***Categorical data***
• Minimum and maximum	• Frequencies
• Mean	- Counts
• Median	- Percentages
• Standard Deviation	• Mode

As we discuss in the visualization section, line charts are more appropriate for metric data, while bar charts are the main choice for categorical data. We use scatterplots to represent the association between two metric variables.

Regression

The workflow for regression modeling is very similar in all cases, regardless of the particular type of application, and it is as follows:

1. Prepare the dataset depending on the type of dependent variable (metric or categorical)
    a. Determine if the models we plan to run need datasets in *wide* or *long* formats.
    b. Building a correlation table or chart is an effective complement to understanding relationships among variables.

F. DePaolis, *Workflows for Climate Change Adaptation Analytics*,
https://doi.org/10.1007/978-3-032-14004-3

   c. Transforming some variables into factors may be necessary to include them as indicator (*dummy*) variables.

2. Build the models (in R, there are different functions depending on the particular type of model we need to build)
3. Verify the significance of the model as a whole (using indicators such as R^2 and *F-test*)
4. Significance of individual independent variables (*t-statistic*, *standard errors*, and *p-value*)
5. Determine the most influential variables (computing the $\beta-$weights)
6. Diagnostic charts to determine if regression assumptions are met.

Time Series

In the Time Series chapter, we discussed alternative forecasting approaches with an increasing level of computational sophistication and analytical complexity. As we move from extrapolation to ARIMA to Bayesian techniques, we apply a common workflow to all of them.

1. Make sure that the observations are equally spaced over time and that the periodicity is known (hours, days, months, etc.)
2. If necessary, we may need to convert the series into a *stationary* one by differencing.
3. Split the dataset into a *training* and a *testing* set.
4. Build the basic models with the training set
5. Test the forecasting accuracy of the model by comparing the forecast from the training set and the testing set.
6. Select the model with the smallest aggregate error (we can use different measures, such as RMSE)
7. Use the best model to forecast values beyond the available sample.

Spatial Analysis

Workflows for spatial analysis are a function of the purpose of the analysis. There are so many different aspects that we cannot describe all of them in a book that has a more general approach. There are, however, some commonalities, and we will focus on those.

1. Data preparation is a fundamental component, especially because we can regularly combine geographic or spatial data with tabular data.

   a. Make sure that the geographic definitions are congruent between data sources (i.e., tabular data at the county level to match with county maps)

   b. Look for discrepancies in the data, such as empty polygons, a mismatch between spatial and geographic data, and misidentification of spatial units (especially when compiling data from different sources)
   c. If there will be a combination of vector data sources, or vector and raster data, make sure that the scales, projection methods, and datum are congruent.

2. Descriptive spatial statistics
3. If applicable, split the data into *training* and *testing* sets for the purpose of validating the forecasting algorithm.
4. Apply appropriate performance metrics to assess the model quality.

A complementary classification of spatial analysis techniques is one based on the equivalence or similarity to those in other sub-fields.

- Equivalent to *Descriptive Statistics and Visualization*, useful in identifying clusters of observations, definition of study areas (based on commonality of features), and many other applications.

  – Visualization of vector data (based on coordinates of geometric units, such as points and polygons)
  – Visualization of raster data (based on a grid of cells or pixels)
  – Construction of DEM (digital elevation models)

- Equivalent to *correlation*

  – Spatial autocorrelation
  – Multiple measures are available, both at a global and local level (Moran's I, Geary's C and Getis-Ord's G)

- Equivalent to *regression*

  – Spatially dependent co-variation, as an extension of spatial autocorrelation for multiple variables
  – Methods such as *GWR* (geographically weighted regression [1]) and *LISA* (local indicators of spatial association [2]) belong in this group.

- Modeling and Spatial Prediction

  – Spatio-temporal covariance functions are essential for modeling dependence in climate data. Interpolation of unobserved values, or *kriging*, is of particular interest in modeling in multiple areas of interest, such as urban planning, environmental and climate science, and epidemiology. For details on random fields and kriging, see Haining [3], Schabenberger and Gotway [4], Chen et al. [5], and Chin et al. [6].

Non-parametrics

Non-parametric methods are intended to temper the strictures imposed by their parametric counterparts. As such, they are more flexible in terms of data requirements. Proper workflows for non-parametric methods are along the lines of determining the most appropriate measure of association between different sub-types of categorical data.

- Purely nominal data
- Nominal data that may have an underlying order
- Strictly ordinal data

In general, and especially in those cases where there are predominant effects, the results of all tests (Chi-squared, Tau, Gamma, etc.) will yield the same results. However, in the relatively few cases where the alignment of concordant and discordant pairs does not point in a clear direction, using the most appropriate test for the type and amount of data in the analysis will better support the analyst's conclusions.

Parting Words

When I first thought about writing a book on this topic, it was not clear to me how I should go about it. Through multiple iterations–too many, perhaps–the final product started to take shape. As it stands now, this book aims to make quantitative analysis more approachable, actionable, and relevant for professionals tasked with solving some of society's most complex and pressing challenges. The workflows introduced here are not intended as final answers, but rather as starting points—templates for rigorous thinking in a dynamic and uncertain world.

Think of these workflows not as static recipes but as living processes. Each project, dataset, and community will demand tweaks, rethinking, and new approaches. The ability to adapt your methods is as important as mastering them. In the context of climate adaptation, answers are rarely final, and decisions are never made in a vacuum. As a practitioner, your role is not just to analyze data, but to help communities navigate uncertainty with insight, transparency, and care.

As you apply these methods, remember that numbers never exist apart from the lives they represent. Data-driven decisions about climate risk must center equity, inclusion, and justice—recognizing that some communities face deeper vulnerabilities yet often have fewer resources.

The field of climate analytics is evolving rapidly, and so are the tools and techniques at our disposal. Whether you're refining workflows, mentoring colleagues, or shaping policy, your capacity to learn—and to share what you've learned—will be part of the solution. As a practitioner, your work is vital. The road ahead will be difficult, but data, thoughtfully applied, can light the way. Use these tools not just to measure risk, but to build hope, resilience, and a more sustainable future.

References

1. Fotheringham, A.S., T.M. Oshan, and Z. Li. 2023. *Multiscale Geographically Weighted Regression: Theory and Practice*. CRC Press. ISBN: 9781000989700.
2. Anselin, Luc. 1995. Local indicators of spatial association—LISA. *Geographical Analysis* 27(2): 93–115. https://doi.org/10.1111/j.1538-4632.1995.tb00338.x.
3. Haining, Robert. 2003. *Spatial Data Analysis: Theory and Practice*. Cambridge University Press. ISBN: 0-521-77437-3.
4. Schabenberger, Oliver, and Carol Gotway. 2005. *Statistical methods for spatial data analysis*. Chapman & Hall.
5. Chen, Wanfang, et al. 2022. *DeepKriging: Spatially Dependent Deep Neural Networks for Spatial Prediction*. arXiv: 2007.11972 [stat.ML]. https://arxiv.org/abs/2007.11972.
6. Chin, Yen-Shiu, Nan-Jung Hsu, and Hsin-Cheng Huang. 2024. Covariate-dependent spatio-temporal covariance models. *Spatial Statistics* 63: 100853. ISSN: 2211-6753. https://doi.org/10.1016/j.spasta.2024.100853. https://www.sciencedirect.com/science/article/pii/S2211675324000447.

Appendix A
R Packages

The following list includes all the packages used in the computer code throughout the book, and it is recommended that you install all of them. Some packages required the installation of other packages that may not be included in this list. Also, notice that some packages, such as kable and stargazer, are used only for formatting tables and the output of some processes.

Complete list of R packages used in the book code

- aod
- bayesforecast
- broom
- caret
- collapse
- corrplot
- data.table
- DescTools
- flextable
- forecast
- foreign
- formatR
- ggplotify
- gt
- gtools
- gtsummary
- Hmisc
- httr
- jsonlite
- jtools
- kableExtra
- knitr
- lubridate
- mapgl
- MASS
- naniar
- pastecs
- plotly
- RColorBrewer
- RCurl
- readr
- rjson
- rms
- rsample
- sf
- sfdep
- spData
- spDataLarge
- spdep
- stargazer
- stR
- tidycensus
- tidymodels
- tidyverse
- tigris
- tmap
- vtable
- xtable

F. DePaolis, *Workflows for Climate Change Adaptation Analytics*,
https://doi.org/10.1007/978-3-032-14004-3